About Island Press

Island Press is the only nonprofit organization in the United States whose principal purpose is the publication of books on environmental issues and natural resource management. We provide solutions-oriented information to professionals, public officials, business and community leaders, and concerned citizens who are shaping responses to environmental problems.

In 2000, Island Press celebrates its sixteenth anniversary as the leading provider of timely and practical books that take a multidisciplinary approach to critical environmental concerns. Our growing list of titles reflects our commitment to bringing the best of an expanding body of literature to the environmental community throughout North America and the world.

Support for Island Press is provided by The Jenifer Altman Foundation, The Bullitt Foundation, The Mary Flagler Cary Charitable Trust, The Nathan Cummings Foundation, The Geraldine R. Dodge Foundation, The Charles Engelhard Foundation, The Ford Foundation, The Vira I. Heinz Endowment, The W. Alton Jones Foundation, The John D. and Catherine T. MacArthur Foundation, The Andrew W. Mellon Foundation, The Charles Stewart Mott Foundation, The Curtis and Edith Munson Foundation, The National Fish and Wildlife Foundation, The National Science Foundation, The New-Land Foundation, The David and Lucile Packard Foundation, The Pew Charitable Trusts, The Surdna Foundation, The Winslow Foundation, and individual donors.

RIVERS
of GOLD

This book is affectionately dedicated
to Moyara, Bert, and Cypress

RIVERS of GOLD

*Designing Markets
to Allocate Water
in California*

BRENT M. HADDAD

ISLAND PRESS

Washington, D.C. ◆ Covelo, California

Library of Congress Cataloging-in-Publication Data
Haddad, Brent M.
    Rivers of gold : designing markets to allocate water in California
/ by Brent M. Haddad.
      p.   cm.
    Originally presented as the author's thesis.
    Includes bibliographical references and index.
    ISBN 1–55963–711–0 (cloth). — ISBN 1–55963–712–9 (paper)
    1. Water-supply—Economic aspects—California. 2. Water supply—
Political aspects—California. 3. Water rights—California.
4. Federal-state controversies—United States. I. Title.
HD1694.C2h23   2000
333.91'009794—dc21   99–27758
                                                                CIP

Printed on recycled, acid-free paper ♻

Manufactured in the United States of America
10 9 8 7 6 5 4 3 2 1

Contents

Preface
A Note on the Author's "Water Politics"

This being a book about water reallocation in a politically charged period, let us move straight to confessions and disclaimers regarding the author's stances on water politics. I grew up in southern California's San Fernando Valley. On clear days, I could see the outflow point of the Los Angeles Aqueduct sparkling on the distant hills. I had no idea where our water came from other than "the pipes." My interest in environmental issues grew out of anger at the fast pace of development in the San Fernando Valley (the transformation of my Little League baseball field into condominiums was the greatest insult) as well as family vacations to Lake Tahoe, where the same kind of development was occurring. My connections to California's farming sector came later on—first as a graduate student and now as a professor in a university department that includes strong agricultural interests. Today, living on California's central coast, I do not have an immediate hydrologic interest in the issues presented in this book. In my community, the water comes from coastal rivers and aquifers and is not connected to the larger state system. Having lived in or visited nearly every corner of the state, I have an appreciation for nearly every interest in the water-reallocation debates. I regularly tell students in my Water Policy classes that the confounding but interesting aspect of water reallocation is that it is a story not of "good guys" and "bad guys" but rather of parties with legitimate and deeply felt interests that, taken as a whole and without compromise, are incompatible.

I approached this topic initially for its connection with markets, as opposed to water (see, for example, Haddad 1997). Before returning to school for my M.B.A. and Ph.D. degrees, I had worked both as a broker of air pollution emission reduction credits and as a securities salesman and

ix

sales manager, licensed to offer stocks, bonds, mutual funds, limited part-
nerships, real estate investment trusts, and many different forms of insurance. I was struck by the different workings of air pollution markets (very
few, costly transactions in those days) and securities markets (numerous
low-cost transactions). My doctoral research was therefore focused more
on markets, with water allocation as a case study, than on water issues with
a focus on market-based reallocation. In short, there was, and is, no political or economic interest driving my perspectives.

I view my role in water politics as informing a debate that others will
decide. The list of design recommendations for water markets in chapter 8
is mine alone, and water-resource professionals will find that nobody is a
complete winner or loser when the list is taken as a whole. If this book
proves useful in informing the water-allocation debate and in providing an
example of how one goes about studying economic institutions and natural resources, I will have partially fulfilled my side of a mutual commitment between myself as recipient of a public education and the public who
paid for it.

How This Book Treats the Subject of Markets

For historical reasons, it has been difficult for writers to separate themselves
from the twentieth century's political polarization with respect to markets.
From roughly 1920 through 1990, countries of the Northern Hemisphere
were divided into two camps, one favoring markets and the other oppos-
ing them. Not only were writers correctly concerned that their comments
about markets could result in political backlashes against themselves, but
also the entire context of the discussion of markets was colored by the
political rivalry between East and West. Only rarely did researchers step
outside the rivalry and address the strengths and weaknesses of markets as
allocation mechanisms. One such example is F. A. Hayek's 1945 article,
"The Use of Knowledge in Society."

Now, a decade after the collapse of the Soviet Union, things have
changed. The "command economy" end of the historical polarity is in
political shambles. But even the "free-market" end of the polarity is not
unscathed. The historical high-water mark for free markets, or "pure markets," came on January 1, 1994, the day the North American Free Trade
Agreement took effect and the Zapatistas launched their rebellion in Chi-
apas, Mexico. The boldness of the Zapatistas, and the broad scope of their
critique of market-based social organization, reminded the world that post-
Soviet-era markets may not solve all social problems and may even create

some. Since then, the inability of some nations, most notably Russia, to adopt market mechanisms suggests that markets have an institutional nature and cultural context that may not be universal.

The world's current agenda with respect to markets has to do in part with how societies can control markets in ways that make the most of their good qualities but minimize their harm. Markets are being questioned on many fronts, including unregulated trading of international currencies, trade in goods made by child and slave labor, trade in goods whose manufacture unduly damages the environment, trade that results in the dislocation or dispersion of indigenous or historical communities, and trade in genetically altered foodstuffs. At the same time, markets are being examined as ways to solve practical problems involving resource allocation. Markets are cropping up in unexpected places and with unexpected products to help subsets of society with allocation problems. The list of novel applications of markets includes air pollution control, fisheries management, reduction of lead in gasoline, and even provision of affordable housing.

Writers interested in markets no longer feel compelled to align themselves along the socialist–capitalist polarity (although some still do by choice). Today, writers can look back over the twentieth century's mountain of neoliberal and Marxian writings on markets like value hunters at a big open-air market of ideas. I am pleased, for example, to have come across A. M. Honoré's analysis of "full or liberal" property rights, which provides an anchor for chapter 3. In sum, this book should be seen as part of the new post-cold-war literature on the institutional nature of markets.

This book is based on the idea that markets in and of themselves have strengths and weaknesses that can be identified, described, and sometimes measured. These qualities can then be compared with those of other resource-allocation mechanisms and a choice among them made. The measure of a market's effectiveness has to do with how consistent the market's qualities are with the details of the context in which it is applied. Market effectiveness can be measured both in comparison with other resource-allocation mechanisms and in relation to minimum standards of social acceptability, ideas that are explained further in appendix 3. The immediate context here is the effort of the state of California to reallocate a portion of its developed water from agricultural use to urban and environmental uses. The larger context is similar efforts in other western states.

This work emerges from the doctoral dissertation I completed in 1996. Compared to the dissertation, there is a great deal of new material here, including this preface, the introduction, chapters 2 and 4, and substantial rewrites of several other chapters. My hope is that interested parties of many backgrounds will read this book comfortably. Indeed, people of many backgrounds make up the state of California and will be facing the choices described here in the near future; if this book proves helpful to them in making those decisions, my effort in creating it will have been rewarded.

Acknowledgments

"Without even knowing it, he had been pulled into that strangest of California sects. He had joined the water people."
—Peter H. King, "Lost Amid the Water People"

Research on this book was launched on February 7, 1994, when Professor Joseph Sax noted to his Water Law class, in which I was a student: "They've been trying to create a water market for years here in California but haven't succeeded. No one knows why." Off I went after the lecture to find out. The search has accompanied me through my graduate-student, post-doctoral, and professorial days.

In getting to answers and, eventually, to recommendations, I have had great help along the way. First, I am indebted to the professors who took an interest in this project both in its dissertation phase at the Energy and Resources Group, University of California, Berkeley, and in its book phase: John Holdren, Richard Norgaard, Jeffrey Romm, Joseph Sax, and Oliver Williamson. I also received advice and comments on drafts from fellow students Andy Cohen, Bill Golove, Rich Hayes, Denny Kelso, Ann Kinzig, Katti Millock, Jurgen Salay, Tom Starrs, Corina Stetiu, and Anne Takahashi. More recently, my students in the Department of Environmental Studies at the University of California, Santa Cruz, including Joan Brunkard, Kimberly Merritt, Karsten Mueller, and Bruce Paton, have commented on drafts of the book and provided research assistance.

When I was getting started, I received valuable assistance in assembling bibliographies on water marketing from Penn Loh, Deborah Moore, Linda Vita-Sunnen and her staff at the University of California's Water Resources Center Archives, and Gary Weatherford. Dave Stoldt also provided valuable insights into the flows of both water and money.

A number of other water-resource professionals went out of their way to assist me in preparing this book. I am especially grateful to Fadi Z. Kamand and Timothy Quinn of the Metropolitan Water District of Southern California; Norman Hill, at that time with the California Department of Water Resources; Gerald Davisson of the Palo Verde Irrigation District; George Core of the Castaic Lake Water Agency; Marc Carpenter and Stephen Ottemoeller of the Westlands Water District; Adrienne Alvord of the Rural Water Impact Network; Gerald Shoaf, who represents the Coachella Valley Water District; and Paul Bartkiewicz, who represents the Yuba County Water Authority. In the highly charged atmosphere of California water politics, these professionals generously assisted me without attempting to assert any control over what this book eventually would say.

I am deeply grateful to the providers of my research and travel grants, including the National Science Foundation, the Energy Foundation, the John D. and Catherine T. MacArthur Foundation, and the University of California Centers for Water and Wildland Resources. I also undertook some field research while I served as a consultant to the Pacific Institute for Studies in Development, Environment, and Security.

To my Island Press editor, Todd Baldwin, thank you for your patience and your faith in this project. And thank you to my extended family, Hans and Maria Ruehsen, for your support throughout. And to my immediate family, Cypress, Bert, and Moyara Ruehsen-Haddad, thanks for your love and patience.

To those I have inadvertently left off this list: please accept both my apologies and my thanks. And to all who have contributed to this research, I hope it returns at least a portion of your investment. That is for you to decide. As William Mulholland once said, "There it is: come and take it."

Introduction
The Ghost of Owens Valley

The Owens Valley lies east of the Sierra Nevada range, completely within California's boundaries, with the White Mountains rising on its Nevada side. The valley's largest town is Bishop, with a population of 3,500. Shortly after the turn of the twentieth century, officials of the city of Los Angeles assembled lists of farm owners in the Owens Valley. Disguising their identities, city agents then approached the farmers individually and offered them option contracts to sell their land—cash payments in exchange for the option to purchase their farms in the future at a specified price. Most of the farmers voluntarily agreed. They signed the contracts. When nearly all the farmers in the valley had signed, the city of Los Angeles exercised its options and took ownership of the farms. When the few remaining farmers began to realize that their land was losing value as a result of the buyouts, the city generously agreed to purchase their farms at the earlier, higher value.

Why is this arguably the most important single event in California's water history? Clearly, Los Angeles did not want to farm the land. It was after the water. The farmers had water rights to the Owens River, which runs north to south down the middle of the valley, and when their land was sold, their water rights accompanied the land. By 1913, after it had built a huge aqueduct from the Owens Valley across the Mojave Desert, Los Angeles diverted what had been water destined for valley farmland to urban uses along California's southern coast.

Yet the importance of the Owens Valley is even greater than this enormous diversion of water from agricultural to urban uses. In fact, when water people get to talking, the very name *Owens Valley* stands by itself as a complete sentence. Why do the events in the Owens Valley still hover like a ghost over water planning, not just in California but throughout the arid West?

The answer to this question lies in similarities between conditions at the turn of the twentieth century and conditions today. That is to say, in its basics, the situation has not appreciably changed. California farmers still use the vast majority of developed water in the state. California still has thirsty, growing, well-financed cities that are searching for new water supplies. Farms were and are an obvious source. Although today there are other voices in water-supply debates—environmentalists, hydropower interests, recreational interests, and regional growth interests—the two most important camps in water debates remain the cities with their water agencies and the farmers with their irrigation districts. If anything, this context is more threatening to farmers today because the state now has adequate infrastructure to move water from almost any farming region to almost any major city. The two key elements of the state's infrastructure are the California Aqueduct, which runs north to south, and the Colorado River Aqueduct, which runs east to west. Both aqueducts end up in the urbanized and rapidly growing southern-coast region. Neither existed in 1905.

Today, the first sentence on the town of Bishop's "Welcome to Bishop, California!" World Wide Web page sums up the effects of the transfer: "Bishop sits directly in the middle of about a zillion square miles of incredible natural beauty. . . ."[1] Today, a beautiful, quiet, empty valley lies behind the Sierra Nevada. The town of Bishop survives, as do a couple of others, but the Owens Valley is no longer the booming region it was at the turn of the twentieth century.[2] The farms are largely gone. With them went farmers and farmworkers and their families; grain stores and implement shops and their owners, employees, and families; and so many other elements of a thriving agricultural community. Today, the Owens Valley is an outdoor recreation destination and little more.

What is the importance today of the Owens Valley story? It is the cautionary tale of what can happen to an agricultural community when farmers surrender water rights to a thirsty, growing city. When the water goes, so goes the way of life.

A New Era of Water Reallocation

Numerous trends suggest that a further significant reallocation of water from agricultural to urban regions is likely to occur in western states in the coming decades. Certainly not all or even most agricultural water will be reallocated, but reallocation of as much as 15 percent of current agricultural usage is plausible. If cities in California were to get 15 percent of what agriculture in California currently uses, urban water availability would rise by more than half. This would be more than enough to meet new urban demands well into the twenty-first century.

An overall reduction of 15 percent or less in agricultural water use may not seem like much to an outsider; it leaves 85 percent of current usage in agriculture. But what if it were *your* irrigation-dependent farm being asked to give up the water, or your entire irrigation district? And what if, as in the Owens Valley, it was not just 15 percent that your region was going to lose, but in fact even more would be needed to preserve ecosystems and their functions? What if your farm and your region were barely getting along economically, but you and your neighbors were strongly committed to rural ways of life and to farming? Why should your way of life be sacrificed to fill swimming pools in southern California or to facilitate urban sprawl? And finally, property rights and contractual rights matter. If farmers and irrigation districts have rights to 85 percent of the state's developed water, why should they not be allowed to keep them if they so choose?

There is a clash of legitimate interests over the topic of water reallocation. Cities point to growth projections that indicate they will need more water to sustain public health and economic growth. Environmentalists point to scientific studies and public-opinion surveys that call for conservation and restoration of degraded waterways and wetlands. And farmers point to the centrality of water to their livelihood, to the cultural preservation of entire rural communities, and to food security for California and the United States.

Since the Owens Valley episode, California and other states have managed to avoid the collision of these three interests simply by developing new supplies. New dams, reservoirs, and aqueducts have been constructed so that additional surface waters could be allocated to cities and irrigation districts. But we are now at a point where new supplies are unlikely to save states from having to make hard reallocation choices. Most of the best dam sites are already occupied. That means that new dams will be more costly and less productive in terms of water storage and hydropower production. New dams will face tighter scrutiny with respect to environmental damage, and their owners will be responsible for costly mitigation. They will probably face opposition from well-organized local and nonlocal groups. Moreover, they will very likely have to pay for themselves, without the benefit of federal grants and loans.

One last factor that has postponed this new era of water reallocation is rapidly disappearing. It is the excess supplies of water historically delivered to cities. These excess supplies gave the impression for decades that Los Angeles and other southern-coast cities were water-rich even though it rarely rained there and no significant natural rivers flowed nearby. Today, however, no one labors under the illusion of excess. Conservation practices have stretched water supplies. While the amount of water that cities can truly conserve is still debated, certainly the reservoir of potential conservation is much lower at the beginning of the twenty-first century than it was even a decade earlier. In sum, the new era of water reallocation has come, with all of the long-postponed painful choices that era entails.

Water Markets: The Reallocation Policy of Choice

In the 1970s, water markets were first introduced as a potential mechanism for reallocating arid-state water resources. Such markets would redistribute water to higher-valued uses: those who could generate value from water equal to or higher than the market price would pay that price, and others would conserve or find substitutes for freshwater. The proposals were grand and optimistic: a statewide California water market would be established in which any water user could buy and sell water rights. California already had the necessary plumbing to provide conveyance mechanisms for water transfers. A prevailing price for water would emerge. Owners of water rights could make good decisions about whether to hold or sell their rights. Water-market advocates identified the many positive aspects of markets in general (such as economic efficiency, individual choice, and political neutrality in reallocation) that would accompany water markets.

Still, sides were soon drawn with respect to who favored and who opposed water markets. Not surprisingly, market advocates primarily included those who wanted to acquire water and had lots of money to do so: urban water agencies. Allied with the urban agencies were banking, industrial, and development interests, all of whom were concerned about the effect of long-term water shortages on urban growth. Pro-market economists also supported water markets, and they lent their analytical powers to policy debates. Environmental interests lent their support because they saw markets both as a way of relieving pressure to build new dams and as a potential new source of water for environmental restoration and conservation.

Market opponents were more diverse and their positions less clear. Farmers and irrigation districts were not opposed to a market mechanism per se, only to one that stood the chance of being dominated by urban interests. Use of short-term markets for water rights (i.e., seasonal or year-long transfers of water) within and sometimes between irrigation districts has been a common practice in agricultural regions for decades and has been acceptable to farmers. Although these opponents of long-term markets did not clearly enunciate what kind of water market would be acceptable to them, in water-market debates, agricultural interests have often distinguished between water as a *commodity* and water as an *input to the agricultural economy.* The distinction lies in both the direction of desired trading and the volume of trading. If water is characterized as a tradable commodity, the policy implication is to create bulk markets that reward the highest bidders, presumably cities and other governmental entities on behalf of environmental interests. But if water is characterized as an input to the agricultural economy, the direction of reallocation is rural to rural and the volume of trading is reduced. This rhetorical distinction is so important to some representatives of agricultural interests that they avoid using such terms as *water market* and *water trading,* instead using the term *water transfer* because it does not necessarily imply a market.

Other water-market opponents were unequivocal, but their lack of organization, small numbers, and lack of financial clout meant that their voices were barely heard in policy debates. In general, anyone who did not control water rights but benefited from their current allocation was likely to oppose market-based reallocation. Officials of rural county and city governments feared that if farms in their jurisdictions were fallowed as a result of water transfers, the local economy would suffer, unemployment would rise, and the demand for public services would rise at the same time that the tax base dwindled. Farmers dependent on well water feared that neighbors who transferred surface-water rights for profit might then tap into the aquifer, thus drawing down a shared resource in order to profit from a private transaction. Finally, parties who wanted water but knew they could not compete on price with urban interests in the southern part of the state also were likely to oppose water markets. This included development interests in the northern part of the state who feared that market-based transfers to southern California could tie up the existing supply (much of which originates in northern snowfall) for decades, thus postponing the north's economic development.

With a strong constituency supporting water markets, and the strongest voices in opposition to water markets merely making the case that some kinds of water markets were acceptable but other kinds were not, California's state legislature took steps in the 1980s and 1990s to establish water markets.[3] Market-based water reallocation became, and remains, the stated policy goal in California.

So Where Is the Market?

Despite twenty years of diligent effort by interest groups and the state legislature, remarkably, no market for water rights has arisen in California.
Numerous long-term interregional rural-to-urban transfers have been
proposed, but few have been completed, and none has actually resulted in
delivery of water to the new owner's service territory. This is a confounding result, given California's long record of pioneering new markets. There
are new markets forming in cyberspace, new markets arising for esoteric
derivative securities, and even new markets developing for illegal goods of
all kinds. Post-Soviet-era Russia might find it difficult to get a market off
the ground, but Californians should not. Yet the record of the last two
decades of the twentieth century is clear: efforts to employ markets as California's mechanism for reallocating its water resources have been diligent
but unfruitful. The reasons are complex and numerous. Mainly, they
involve a failure of most proposed market mechanisms to take into account
the unique historical, political, legal, economic, and cultural context of
California. Contrary to popular belief, markets are not a one-size-fits-all
solution.

Institutional Analysis of Water Markets

Most analytical works on water markets fall into one of two categories:
those that determine how to achieve economic efficiency (or simply state
that it would be a good thing to do so) and those that explore the unintended consequences of the pursuit of economic efficiency. Economic efficiency is a theoretical state of affairs in which, through free trade, all parties are made as well off as they possibly can be, given their initial wealth.
This book recognizes the importance of economic efficiency as a theoretical goal but notes that as long as a theoretical state of economic efficiency
has been the end point of policy efforts to reallocate water, those policy
efforts have been ineffective. In this book, the merits of economic efficiency serve more as a guide than as an end point, with the focus instead
on identifying incremental improvements in policy. Many of the long-term
water challenges in California and other states can be adequately addressed
by creating institutions that are appropriate and feasible but will not bring
about a perfectly efficient market. The compromises in institutional design
that prevent a market mechanism from redistributing water in an economically efficient way may be the very qualities that make it effective. For
example, as described in chapter 2, efficient markets require that water
rights be concentrated in individual buyers and sellers. But doing so also

strips other parties who share elements of the water rights (e.g., pertaining to environmental protection or to the rights of irrigation districts) of their historical claims to the water, as detailed in chapter 3. These other parties have shown that they have the power to block the institutional changes that would concentrate water rights in individuals. Thus, when an economically efficient market is viewed as the sole end point of policy formation, policy gridlock can result.

New avenues for thinking about market formation present themselves when efficiency is viewed as a guide to incremental improvement. The new avenues are no less analytically rigorous than the old, and they still fall within the province of economics, but they are different. The basic challenge is to identify institutions that are consistent with the needs of water traders in California's multifaceted water-reallocation context. What water traders want are institutions that allow them to engage in low-cost, reliable transactions. For example, sellers want assurance of payment and buyers want assurance of delivery. Further, buyers want low-cost means of ensuring the seller's performance (fallowing, crop substitution, etc.). So far, these concerns are consistent with an efficiency-based approach.

But with respect to context, farmers and other rural dwellers want assurances that the sale of water to a city is not going to make their entire agricultural region decline or collapse. They fear that a market, possibly through a series of otherwise independent transactions, could result in a repeat of the Owens Valley story. Similarly, environmentalists want assurances that water transfers will not dry up wetlands, damage natural conveyance facilities, or cause other loss of wild-species habitat. People in all parts of the state want assurances that water markets will not fuel suburban sprawl where attractive open spaces exist.

Accommodating the ghost of Owens Valley and other issues is a matter of institutional design. Getting to design principles is a matter of identifying the goals of water reallocation; examining the historical record of water reallocation; identifying reallocation mechanisms that have worked in the past, and why; and generalizing those mechanisms for broader application. For example, in chapters 5–7, a transaction-cost approach is taken in an effort to demonstrate what made some transfers work where so many others have failed. Those chapters, each a case study of a successful long-term transfer, view the signing of the water-transfer agreement as a theoretical turning point at which arm's-length negotiation changes into joint implementation. Events leading up to the signing are important because they identify how the parties came to grips with external constraints of their agreement in cost-minimizing ways. Events following execution of the agreement offer insights into how the parties adapted their agreements during implementation as unexpected challenges arose. These cost-minimizing adaptations to context before and after execution of the agreements—what they were and why they worked—provide the insights necessary to design effective water-transfer institutions. Finally, to arrive at recommendations, this book reviews the context of water trading in California (and, to a lesser extent, in other western states), a proposed market mechanism, and the experience of successful utilization of short-run market mechanisms in California.

The Plan of This Book

This book takes a new look at California's water-reallocation challenge. Chapter 1 begins by providing an overview of the state's water-supply challenge. California's situation with respect to market reallocation is compared with that of other western states. Next, chapters 2 and 3 examine the market mechanism both in theory and in its specifics with respect to the political processes and legal regime governing California's water. The conclusion of these two chapters is that the original grand vision of a statewide water market was just too grand. The appropriateness and feasibility of a market mechanism can be determined only by looking at the details. That is, markets are complex sets of institutions, expectations, and patterns of behavior. Every market is unique, and to be effective, it must be consistent with its context—the historical, political, legal, economic, and cultural experience of the people affected by the market as well as the underlying demographic, ecological, and physical conditions of the region. In California, all these aspects of context influence the very practical question of whether an urban water agency or environmental interest will be able to secure water rights through a market.

The book then engages the details of California's water-market experience in both short-term regional trading (chapter 4) and long-term interregional trading (chapters 5–7). These chapters identify the extent to which water markets can be effective in California and why, and draw general lessons from California's experience with water markets.

Chapter 8 presents a list of design recommendations for water markets. These recommendations, if followed, could break policy logjams over water reallocation in California and elsewhere. They are based directly on the experience of water marketing in California as examined through the lens of institutional analysis. These recommendations offer both an agenda for change and, more broadly, a way of thinking about the reallocation of scarce natural resources.

This institutional approach to evaluating market-based water reallocation in California is clearly sensitive to the unique context in which the market would operate. Although, as stated earlier, every market (and context) is unique, many of the insights presented here are transferable. The most obvious regions of application are other states in the arid western United States. Beyond that, water-resource professionals in any region that combines aridity with a market economy might find this book useful. More broadly, individuals with a general interest in applying institutional analysis to resource-management problems might also find this book valuable.

Notes

1. Internet: http://www.thesierraweb.com/bishop/default.html, August 1999.
2. The first sentence of the Web site of another surviving town, Lone Pine, reads "Sure, the town's only got one stop light." Internet: http://www.thesierraclubweb.com/lonepine, August 1999.
3. The appendixes to this book chronicle both the efforts to create water markets in California and the results of those efforts.

CHAPTER 1

The Water-Reallocation Challenge in California and the West

"[Between 1995 and 2020,] California's population is forecast to increase by more than 15 million people, the equivalent of adding the present populations of Arizona, Nevada, Oregon, Idaho, Montana, Wyoming, New Mexico, and Utah."

—1998 *California Water Plan Update* (DWR 1998a, p. ES1–4)

A discussion of water reallocation begins with four questions: How much water is available? Where is it now? Who needs it? How will it get from those who have it to those who need it? In answering these questions for California, our starting point is an overview of the state's evolving circumstances of hydrology and demography as well as the legal doctrines that define who has water rights and what the rights entail. Historical, political, social, economic, and ecological perspectives are then woven in to develop a fuller picture of the state's water-supply challenges. After California's features are discussed, this chapter presents a comparative perspective, describing conditions in four other western states: Arizona, New Mexico, Texas, and Colorado.

Water Reallocation: A Key Issue in California

Unlike the situation in some parts of the United States and the world, the water challenges facing California do not emerge from a basic scarcity of the resource. An average of 200 million acre-feet (maf) of precipitation is deposited in California annually.[1] Some regions receive as little as 4 inches per year, while others receive more than 60 inches. The state also has historically appropriated roughly 7 maf of water from the Colorado River and the Klamath River. Given California's population of 33 million, this quantity of annually renewing water is more than four times what some analysts believe is a minimum for agricultural self-sufficiency. In fact, California is one of the world's leading exporters of agricultural commodities, and this performance is fueled in part by the abundance of water.

Of the state's precipitation, two-thirds is consumed through evaporation and transpiration by plants. The remaining one-third comprises the state's annual runoff of about 71 maf. The state also takes advantage of vast groundwater reserves, in an average year withdrawing about 14 maf of the estimated 850 maf available. These natural storage basins supplement California's water during the dry months and during drought years. Since the groundwater reserves are replenished at a rate of 12.5 maf/year, the California Department of Water Resources (DWR) estimates an annual overdraft of 1.5 maf of groundwater. In other words, Californians are mining groundwater faster than it is being recharged by precipitation. A small but growing additional year-round source of water is urban reclaimed water, which is expected to provide a reliable annual supply of almost 600,000

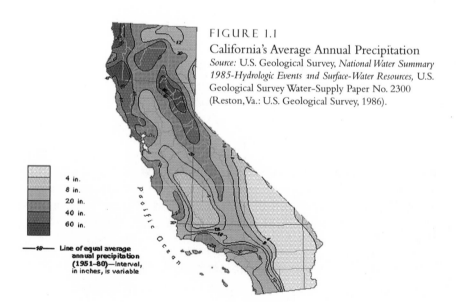

FIGURE 1.1

California's Average Annual Precipitation

Source: U.S. Geological Survey, National Water Summary 1985-Hydrologic Events and Surface-Water Resources, U.S. Geological Survey Water-Supply Paper No. 2300 (Reston, Va.: U.S. Geological Survey, 1986).

4 in.
8 in.
20 in.
40 in.
60 in.

—10— Line of equal average annual precipitation (1951–80)—Interval, in inches, is variable acre-feet (af) for urban and agricultural uses by 2020. A portion of total state supply goes to the maintenance of healthy rivers, estuaries, and wetlands. In average precipitation years, the flow utilized for ecosystem maintenance is 36.9 maf, whereas in drought years the amount used drops as low as 21.2 maf. Thus, in an average year, roughly 42.6 maf is available for agricultural and urban uses. Agricultural production currently uses roughly 80 percent of the available 42.6 maf.

As shown in figure 1.1, most of California's precipitation falls in the eastern and northern parts of the state. A portion of the precipitation, much of which arrives as winter snow, is captured in reservoirs during the spring thaw and then released throughout the dry, hot summer and autumn months. A combination of dams, aqueducts, and pumping stations—constituting the most developed water system in the world—augments natural rivers and delivers water from the mountains to Central Valley agricultural regions and coastal population centers.

Changing Patterns of Water Demand in California

As in most of the western United States, water in California was originally allocated to serve an agrarian economy. The federal government's primary mission was to "reclaim" wildlands and create opportunities for westward migration, homesteading, expansion of agriculture, and economic growth. As a result of this pattern of expansion, economic and political power came to reside in rural areas, not cities. Irrigation districts and individual farmers locked up the lion's share of California's developed water, in the form of either renewable long-term contracts or permanent rights.

As have other western states, California has been undergoing a transition in its use of freshwater resources. Net water use in the early 1990s included 80 percent for agriculture, 16 percent for urban regions, and 4 percent for recreation, wildlife, and power generation (MacDonald 1993). The largest agricultural users in terms of acreage were cotton (1.5 million acres), irrigated pasture and alfalfa (1 million acres each), and rice (0.5 million acres). Water use in the state's agricultural regions peaked around 1980 and has been slowly declining since then. Statewide demand continues to grow but at a historically low rate, bolstered by demand from urban regions, which is growing at a rate of roughly 64,000 af/year.

Both aggregate demand and the pattern of use are expected to change in California as population grows from 1995's roughly 32 million to the

47.5 million projected for 2020 (DWR 1998b). Figure 1.2 highlights projected regional trends in population growth. When these trends are compared with the distribution of rainfall shown in figure 1.1, California's water-transfer challenge stands out in stark relief.

Underscoring the urban–suburban focus of California's future water needs, in 1995, 110 new towns and major subdivisions were in the planning stages. These were expected to accommodate 2 million new residents; 30,000 acres of commercial, industrial, and office space; and at least forty-four golf courses (Arax 1995). These projects will require 0.6 maf of water on a long-term basis.

Just as growing populations require long-term rights to water, industry does as well. Industry uses water to cool energy-generation equipment, for cleaning purposes, as inputs to production, and for consumption by employees. A firm evaluating alternative sites for a multimillion-dollar factory may make the certainty that sufficient water will be available over many decades a decision criterion. Long-term rural-to-urban transfers therefore are consistent with the long-term water-use needs of a state's growing industrial sector (Kay 1994). The fact that industry requires a consistent, reliable year-round supply of water stands in contrast to California's patterns of seasonal precipitation and runoff and to the changing seasonal demands of much of California's farming sector.

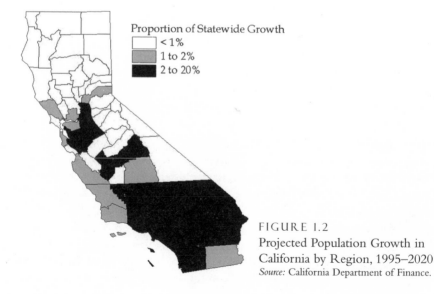

Proportion of Statewide Growth
☐ < 1%
▨ 1 to 2%
■ 2 to 20%

FIGURE 1.2
Projected Population Growth in California by Region, 1995–2020
Source: California Department of Finance.

Environmental needs for new water resources are also growing. Efforts to restore salmon runs on the San Joaquin River and other rivers require dedicated instream flows, even when the state is experiencing a drought. Efforts to restore wetlands require new appropriations for proposed wetland regions, and efforts to restore the historical seawater–freshwater balance in the San Francisco Bay/Sacramento–San Joaquin Delta (Bay-Delta) region require new dedicated flows of freshwater as well.

The 2000–2020 scenario that emerges is one of California's agriculture giving up water (losing 2.3 maf, or 7 percent of current annual usage), urban areas acquiring water (gaining 3.2 maf, or 36 percent over current annual usage), and environmental uses increasing by about 100,000 af/year (a 0.1 percent increase) (DWR 1998b). The resulting anticipated annual shortage of 1.0 maf may be compounded by the loss of as much as 900,000 af/year of Colorado River water, by the possibility of additional water being dedicated to the Bay-Delta ecosystem, and by efforts to end the overdrafting of California's aquifers. If current patterns of demand and use continue in California, with its existing facilities, by 2020 the state could fall short of meeting its water needs by 2.4 maf in normal years and by 6.2 maf in drought years (DWR 1998a).

Location of water use is expected to shift from northern and central agricultural regions to the southern coast, consistent with projected regional demographic growth. Transfers of water from north to south are not new to California. Millions of acre-feet of water are already transferred from the northern and eastern mountains, where water arrives as rainfall and snow, to the central and coastal areas. Growth in demand is expected to be highest in the southern coastal region (0.9 maf, or a 17 percent increase over current annual usage), and the neighboring South Lahontan region (0.3 maf, or a 45 percent increase over current usage). This scenario of change is consistent with demographic growth in which urban population density grows and new suburbs are attached to existing urban areas on the southern coast.

One indicator of how demographic change in the twentieth century favored California's urban sector over its rural sector is the number of state legislative representatives representing rural versus urban regions. In 1902, roughly 25 percent of California's forty state senators represented primarily urban districts. By 1990, that figure had grown to roughly 67 percent. Similarly, between 1902 and 1990, the percentage of state assembly members representing primarily urban districts rose from 35 percent to 70 percent. This long-term trend toward urban concentration seems to be continuing in California. Out of fifty-eight counties, the five largest in terms of population, all urbanized and located along or near the Pacific coast, absorbed 2.9 million new residents between 1980 and 1990, or 47 percent of the state's total population growth.[2] Los Angeles and Orange Counties alone grew by nearly 2 million people. With the trend shifting toward more support for urban interests in legislative bodies, the farm sector is that much further challenged to maintain its historical control over the bulk of the state's developed water.

The Legal Context of Water Reallocation

Legally, there do not appear to be strong barriers to changes in current patterns of water ownership and use. California's system of water law is more complicated than those of most states because it recognizes the water rights of both riparian users (owners of land adjacent to waterways) and appropriators (parties who have built conveyance structures from waterways to the place of use). Most states recognize only one type of right. Under either doctrine, the easiest way to transfer water rights is to transfer land rights and include the water rights in the deal. This is, in fact, the only way in which riparian water rights can be transferred, since riparian rights are considered to be appurtenant to, or inseparable from, the land. Appropriators, however, usually have the option of transferring ownership and changing the point of diversion from the waterway, the timing of diversion, the point of return flow, and the use of the water. Long-distance transfers from one water basin to another also are permitted. That means that a water market in California could include agricultural-to-urban transfers, but it would be a market among appropriators or those who contract with appropriators for water deliveries.[3] It would not involve riparian rights.

The two key principles of the appropriative doctrine are "first in time is first in right" and "no harm." The earliest appropriators get to satisfy their established use patterns first, even in a drought year, while those who began appropriating later wait their turn.[4] During a drought, the most junior (most recently established) appropriators may not be able to divert any water at all. When a water right is transferred, it retains its ranking in comparison with those of other appropriators on the same waterway. But any action taken must not harm the established rights of other appropriators (as well, in California, as riparian users). That means that new diversions are not allowed on waterways that are fully appropriated, and a water transfer would not be allowed if it harmed the ability of other appropriators to utilize their rights.

The "no-harm" portion of the appropriative doctrine protects only those parties who hold water rights. The rights of other interests—environmental protection, rural towns, recreation, farmland, or open-space preservation—are protected by other statutes, such as that requiring an environmental impact statement for large-scale transfers. Although the law contains no centralized statement of the rights of nonappropriators and non–riparian users with respect to water transfers, both formal and informal avenues of recourse exist, and many proposed transfers have been blocked in part by parties who were not formal holders of water rights.

Ten states recognize both riparian and appropriative rights,[5] and nine others recognize only appropriative rights.[6] Nearly all of the states that recognize appropriative rights are located in the arid western part of the country, where diversion of river water over long distances has been a central element of economic development for the past 150 years. States that recognize only the appropriative doctrine have an easier time establishing water markets because both water-availability calculations and no-harm considerations are simplified by the absence of riparian rights.

Thus, California water law retains the flexibility to transfer both water rights established under the appropriative doctrine, as well as to recognize contracts for delivery of appropriative-rights water.[7] Irrigation districts and urban water agencies that procure water from major water projects, such as California's State Water Project and the federal Central Valley Project[8] (also located in California), have long-term contracts for water deliveries from those projects. The projects' owners and managers, the DWR and the U.S. Department of the Interior, respectively, hold the appropriative rights, which are not likely to be transferred. Provisions exist, however, for contract holders to transfer contracts both among themselves and to outside parties. Although these contracts are not as strong as actual appropriative rights, they represent long-term commitments by state and federal agencies to deliver water to the contract holders, and at least one permanent transfer of a contract has occurred and a number have been proposed. In normal years, nearly 10 maf of water-project water is delivered, about one-quarter of all developed surface water in the state.

The Water-Reallocation Context in Western States

The economic importance of the agricultural sector in comparison with that of industry, commerce, and government was in decline in western states throughout the twentieth century. Figures 1.3 and 1.4 chart recent

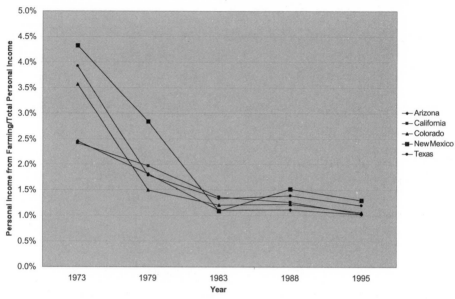

FIGURE 1.3

Proportion of Jobs on Farms in Selected Western States, 1973–1995

Source: U.S. Department of Commerce, 1998.

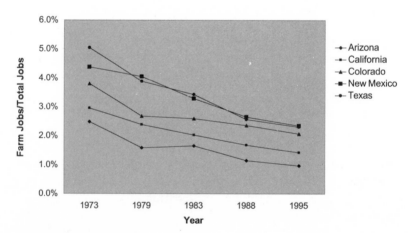

FIGURE 1.4

Proportion of Personal Income from Farms in Selected Western States, 1973–1995

Source: U.S. Department of Commerce, 1998.

declines in the proportion of all jobs on farms and the proportion of all personal income earned on farms for Arizona, California, Colorado, New Mexico, and Texas. Since 1973, both indices have been in steady decline in every state.[9] The proportions of personal income and jobs from farming are startlingly small. Of the five western states listed, in 1995 New Mexico had the highest percentage of total personal income from farming: less than 1.5 percent. California, Colorado, and Arizona each had just 1.0 percent of personal income generated on farms; in California's case, this represented a 57 percent decline over a two-decade period. In terms of jobs on farms, New Mexico and Texas had the highest proportions, 2.3 percent, whereas California had only 1.4 percent of its jobs on farms. Inclusion of downstream agriculture-related industries (statistically described as "agricultural services" and "food and kindred products") roughly doubles the state's proportion of personal income from agriculture, to the 2–3 percent range, but has less of an effect on job figures. In absolute terms, the total number of farm jobs has decreased in California by more than 15 percent between 1973 and 1995, and total personal income from farming has fallen by more than 10 percent. In short, the economic importance of farming in the semiarid West shrank dramatically during the twentieth century.

Meanwhile, urban populations in western states have exploded and are expected to grow further. According to the 1990 census, 86 percent of all residents of western states lived in urban areas, compared with a national figure of 75 percent. California's urban proportion, 93 percent, is the highest of any state in the country. Overall population levels are also expected to rise in the coming decades.

Environmentally, growing public interest in preserving and restoring free-flowing rivers, wetlands, and other wildland areas means that a relatively new demand has been placed on water resources: protection and enhancement of ecosystems. Although California's dramatic loss of more than 90 percent of its original wetlands is the country's highest, Arizona, New Mexico, Colorado, and Texas each have lost one-third to one-half of their original wetlands, also significant proportions.

California is therefore not alone in either its background circumstances or its policy choice to pursue market-based reallocation. Like California, other western states have explored markets as mechanisms for reallocating water. The sections that follow contain brief reviews of the fundamental features of four other western states with respect to water reallocation.[10]

Arizona

An appropriative-doctrine state, Arizona has a population of 4.6 million people, the vast majority of whom live in cities. The price of water in Arizona's two main sectors of end use—urban and agricultural—is strikingly different. Farmers pay $3 to $10 per acre-foot for irrigation water and cities pay $100 to $120 per acre-foot. In an effort to secure water resources, many Arizona cities, including Tucson, have engaged in a practice known as water ranching—purchasing and then fallowing farms in order to secure their water rights. Nearly $200 million was spent in procuring tens of thousands of acres of land for this purpose during the 1980s.

Arizona annually receives only nine to ten inches of rain on average, relying instead on groundwater and Colorado River appropriations. Both of these sources have been enveloped in controversy for years. Disputes over groundwater have centered on the need to maintain the health of aquifers in light of growing urban demand for groundwater supplies; Colorado River concerns have centered primarily on the relative shares of California and Arizona in the river's annual flows. The two states' shares have been contested as far as the United States Supreme Court.

These historical controversies have hindered Arizona from establishing a water market in part because it has been unclear just how much water in-state users have rights to. Currently, the multibillion-dollar Central Arizona Project, a federally funded canal that moves Colorado River water east to agricultural regions in central Arizona, provides plenty of water, but the water is too expensive for the farmers. The canal has the capacity to deliver 2.2 maf/year but is expected to deliver approximately 1.5 maf/year.

Arizona's aquifers have been overdrafted by roughly one-third of their capacity, and this has prompted strict regulation of groundwater use. Since surface-water law and groundwater law are distinct in Arizona, any effort to form a water market in this state will include the challenging task of reconciling the two separate bodies of law.

Another issue likely to be addressed in any effort to reform water-allocation practices in Arizona involves the claims of Indian tribes. Twenty-eight percent of Arizona is tribal land, and the federal government reserves water for tribal use. The same 1963 Supreme Court decision that solidified Arizona's claim to Colorado River water vis-à-vis California also set aside nearly 1 maf for the five tribes of the Colorado River Indian reservations. This amounts to roughly 13 percent of the total allocation available to California, Nevada, and Arizona and disproportionately affects Arizona's share because of the prevalence of tribal lands in that state. The question of tribal water rights will probably be addressed in any serious attempt to create a market for water in Arizona.

One final issue involves efforts to restore wildlife habitat and reduce Colorado River salinity problems. The Endangered Species Act of 1973 and the Clean Water Act of 1977, as well as treaties with Mexico, may push the federal government and others to invest money in water transfers for environmental uses. In sum, although individual efficiency-enhancing transfers of water can be proposed, any statewide reform would face the immense challenges of reconciling Colorado River law, Indian rights, groundwater doctrine, and environmental requirements.

New Mexico

New Mexico is a relatively large state whose population of 1.8 million correctly suggests significant open space. It is another appropriative-doctrine state. At least two current controversies there suggest that water transfers could help resolve conflicts over allocation. In one case, the Ute Mountain Ute Indian Tribe and the Southern Ute Indian Tribe are asking that an engineered water project be built in southern Colorado (the Animas-La Plata project) to deliver water to tribal lands in quantities that would satisfy their rights. However, federal efforts to build the project would very likely encounter a roadblock in the Endangered Species Act. The federal government is offering to purchase back from the tribe their water rights and thereby avoid building the project altogether. The Utes, however, continue to prefer the project to the money.

A second recent water-transfer controversy involves New Mexico's historical overuse of the Pecos River before it enters Texas. In 1987, the U.S. Supreme Court mandated a reduction of use by New Mexicans. In response, the state is attempting to purchase and retire water rights to the river so that sufficient flow crosses into Texas. As of 1991, the total cost to the state to procure Pecos River water rights was estimated to be as much as $60 million (Sax, Abrams, and Thompson 1991). Absent willing-seller transactions, New Mexico will use a queuing approach to water reduction, shutting off access to the most recent appropriators. Both of these cases involve relatively well defined needs for reallocation of water; the federal and state governments are examining options for resolving the problems.

Texas

Like California, Texas is a hybrid-doctrine state that recognizes both riparian and appropriative water rights. Intersectoral competition for water, either by market, political, or judicial means, is growing (Griffin and Ellis 1997). The state's 1990 population of 17 million is expected to double by 2040. Current water use of roughly 16.5 maf/year is concentrated in the agricultural sector (65 percent), followed by the municipal sector (20 percent) and the manufacturing sector (9 percent). The amount of water demanded is expected to increase by 10 percent by 2040, but within the overall growth of water demand, sectoral uses will swing widely. Agriculture's share of total water use is expected to decline from 65 percent to roughly 45 percent, and municipal use is expected to increase from 20 percent to more than 30 percent of the total.

In southern Texas, water markets have served as an effective reallocation mechanism for nearly a quarter of a century. In the lower Rio Grande valley, following a 1970 court-ordered adjudication of water rights, a robust market for water rights and leases has developed. The lower Rio Grande valley provides good conditions for trading, and water rights are now well defined. Further, the possibility of negative third-party consequences arising from a water transfer is minimized by the fact that water diverted from the river is considered to be completely consumed because once it is removed, it evaporates quickly and drains away from the river. The absence of return flow minimizes the possibility of multiple uses of the same appropriation.

Outside the Rio Grande valley, most surface-water markets have been thin because of weak enforcement of water rights, easy availability of groundwater, a single administrative authority supplying all water needs, and an adequate water supply (Griffin and Ellis 1997). Water has been traded within irrigation districts in Texas, and a state-sponsored water bank was established in 1993 to help buyers and sellers trade water. As of 1997, the Texas Water Bank had not yet recorded a transaction, but recent legislative reforms could increase its attractiveness to future buyers and sellers.

Some groundwater has been traded in recent years. Groundwater supplies nearly two-thirds of all agricultural irrigation water in the state. Griffin and Ellis (1997) point out that in early 1997, a San Antonio–based developer accepted the bids of forty farmers who agreed to forgo irrigation on a total of 10,000 acres of land. The average participant was to receive $231 per acre for changing to dryland farming. This pilot project was scheduled to conclude in 1998.

Texas, like California, is experiencing population growth and is projected to grow further. Both states also project a shift in water use from rural to urban areas. Both states have pockets of successful trading but not an effective statewide mechanism for moving water. Texans appear not to be strongly averse to groundwater trading, whereas in California, farmers draw a "line in the soil" when it comes to selling groundwater. If predictions of future water demand and water use by sector are even somewhat accurate for either state, new mechanisms will be needed to move water rights from agriculture to cities.

Colorado

Colorado, informally known as the "headwaters state," is not arid. A flow of 14 maf/year comes primarily from the state's Rocky Mountain snowmelt. Three major rivers originate in Colorado: the Colorado River, which departs the state with a flow of 10 maf/year; the Missouri River, with 2 maf/year; and the Rio Grande, with 1.5 maf/year. Agriculture accounts for 85–90 percent of in-state water use, with a corridor of cities on the eastern slope of the Rocky Mountains, including Fort Collins, Denver, Colorado Springs, and Pueblo, accounting for the bulk of urban use.

Water rights in Colorado are considered to be like other property rights. Colorado was one of the first western states to simplify its system of water ownership, having abandoned the riparian doctrine more than a century ago. Only the prior-appropriation doctrine is recognized. Knowledge of how to negotiate and transfer water rights resides in the private sector, among attorneys and engineers who help negotiate and execute transfers. The role of the engineer is to verify the historical-use claims made by the seller and to evaluate whether the transfer will injure other appropriators along the river. "Injury" encompasses reduction in the amount available for appropriation, costly lowering of stream elevation, increased mineral load, and similar adverse results of water transfers.

Water rights have been traded for years in Colorado. Seven regional offices of the Colorado Division of Water Resources administer hydrologically defined trading regions. Each office is equipped with both engineering experts to evaluate whether injury has occurred and a water court that rules on whether a transfer can go forward. One region, Water Division 5, which encompasses the Colorado River headwaters, administers 22,000 water rights located in 11,000 diversion structures (dams, canals, reservoirs). During the mid-1990s, Water Division 5's regional office was receiving five to fifteen applications to change water rights each month. Prices ranged from $74 to thousands of dollars per acre-foot, in the latter case to provide water to a rapidly expanding area called Summit City. Quantities sold typically ranged from 0.25 to 50 af per transaction.

The comparatively strong property rights granted for water in Colorado mean that the state's public and private environmental interests typically must compete on equal footing with other water users for water supplies for wetlands, instream flows, and other ecological uses. The Colorado Division of Water Resources treats all buyers and sellers alike, be they governmental, public nongovernmental, or private parties. Of course, environmental interests, which lack both profit-oriented financial resources and general powers of taxation, are at a disadvantage compared with better-financed urban and agricultural interests.

Another stress on Colorado's water-trading system is the fact that the urban areas, where growth is spurring increased demand for water, lie primarily on the eastern, drier side of the Continental Divide, whereas the wetter (snowier) regions are on the western side. In addition to an institutional structure not designed for interbasin transfers (each basin, or division, is primarily organized to handle intraregion transfers, which constitute the vast majority of all transfers), state law requires that compensating reservoirs be built whenever water is moved off the western slope of the Rockies to serve the more populous eastern slope. This additional requirement increases the cost of market-based interbasin transfers and reduces the likelihood that they will form a central part of future reallocation plans.

Most of Colorado's trading takes place within the hydrologically and administratively separated water basins, with occasional efforts to move water between basins. Although Colorado differs from California in the relative simplicity of its water-rights structure, in its hydrologically defined administrative regions, and in its role as a major source of water for numerous western states and Mexico, it is similar to California in that its institutional structure was designed neither to move great quantities of water to growing urban areas nor to accommodate new environmental demands for water consistent with the changing social values of the past few decades.

Meeting the Water-Reallocation Challenge

Nearly every drop of water in the arid West already has a claimant, and sometimes multiple claimants. At the same time, total demand for water is growing in absolute terms as arid-region populations grow and as society expresses increased interest in preserving and restoring natural systems. The pattern of demand also is changing as the arid West urbanizes, as the economic importance of agriculture declines, and as calls are made for increased environmental protection and restoration.

Where will new supplies come from? Water once considered surplus and available (free-flowing rivers, for example) is now recognized as playing an important ecological role in maintaining rivers, estuaries, and riparian zones. Dams and reservoirs that capture spring floodwaters themselves produce adverse environmental effects, making the possibility of major new projects remote.

Thus, if one interprets the question "How much water is available?" as referring to *new* water, a fair answer is "Essentially none."[11] Nearly every drop that finds its way into surface flows or resides in usable aquifers is connected formally or informally to claims on existing uses. But if "available" water includes water already put to use, a significant amount is available—tens of millions of acre-feet. This is, in fact, the current understanding of the term *available water*. It implies that new demand can be met only with *reallocations*. This in turn means that some proportion of current claimants—the vast majority of whom are in the agricultural sector—are likely to see their use rights and use patterns give way to new claimants, likely from urban regions or representing environmental interests.

The remaining question is "How will it happen?" It is one thing to say that a legal basis for water transfers exists in the appropriative-rights doctrine. It is a far more complicated issue to establish a new set of social institutions for reallocating water that is feasible (i.e., gets the reallocation job done) and appropriate (i.e., is consistent with the vast range of legal precedent, social values, and historical context). Our existing water-ownership and water-management institutions are designed to meet growing demand with growing supply. In their current form, these institutions are neither designed for nor capable of meeting the new and very different challenge of water reallocation. For example, water districts are prohibited from setting a price for water that would reduce demand if that price results in the district earning more money than it cost to provide the water. With strong price mechanisms unavailable to curb individual demand, finding new supplies becomes an obvious alternative. Institutional change is urgently needed, and the process of selecting and establishing new institutions is already under way.

Immense academic, policy, legislative, and lobbying energy has been devoted to the establishment of water markets in California (see appendix 1) as well as in other western states. However, the kind of market a person on the street might recognize does not yet exist for long-term interregional transfers of water. States are still in an exploratory and experimental phase of policy development.

Because the institutions themselves—the rules and procedures that will govern water reallocation—are certain to have both direct and indirect effects on the outcomes of reallocation (who gives up water, who gets it, how much, and under what conditions, to name a few), it is timely and important to carefully examine water markets: What are they? How do they work? What would they look like? This new set of questions—all related to the challenge of reallocating freshwater resources in the coming decades—occupies the next chapter of this book.

Notes

1. One acre-foot of water covers an acre of land to a depth of one foot. Many crops require irrigation at the rate of approximately three acre-feet per acre per year. Alternatively, one acre-foot of water will sustain two to four families for one year.
2. The counties are Alameda, Los Angeles, Orange, San Diego, and Santa Clara.
3. Potential participants in water markets are irrigation districts, and perhaps even individual farmers within the irrigation districts, that contract with such appropriators as the federal Central Valley Project and the State Water Project, among others.
4. An interesting water-marketing approach using the appropriative doctrine is for a junior appropriator to pay a senior appropriator to leave some or all of his or her allowed quantity in the river, which then becomes available to the junior. No water rights change hands. Such changes in water-use behavior transfer water without disturbing the existing rights structure. This approach was utilized in two of California's three leading examples of water transfers (see chapters 5 and 6).
5. In addition to California, nine other states—Kansas, Mississippi, Nebraska, North Dakota, Oklahoma, Oregon, South Dakota, Texas, and Washington—recognize both riparian and appropriative rights. This approach to water rights is known as either the hybrid doctrine or the California doctrine.
6. Appropriation-only states include Alaska, Arizona, Colorado, Idaho, Montana, Nevada, New Mexico, Utah, and Wyoming.
7. In the interest of brevity, this overview of water rights does not cover such topics as federal reserve rights, pueblo rights, pre- versus post-1913 appropriative rights, groundwater rights, and the effect of stream adjudication on water-transfer rights. Water law texts, such as Sax, Abrams, and Thompson 1991, provide thorough coverage of these topics.
8. The State Water Project and the Central Valley Project appropriate and deliver water to cities and irrigation districts throughout California. They include the state's biggest dams and longest aqueducts and canals as well as

numerous pumping stations, flood–control facilities, and environmental-protection facilities. Together, the two projects deliver nearly 10 maf in a normal water year.

9. New Mexico showed a brief rise in its population of jobs on farms in the early 1980s in part as a result of the savings and loan crisis.

10. A primary source of information for this section was the workshop "Economics of Water Marketing in the American West," held July 17, 1997, at Berkeley, California, and sponsored by the Department of Environmental Sciences, Policy, and Management of the University of California, Berkeley, and the Environmental Protection Agency. Presenters included Bonnie Colby (Arizona), Micha Gisser (New Mexico), Ron Griffin (Texas), Bill Boggess (Oregon), Janis Carey and Richard Howitt (California), Ron Johnson (Montana), Bob Young (Colorado), Gary Green (Washington), and Dave Sunding (Nevada).

11. Recycling and desalination present limited and expensive new–water alternatives that can increase supplies in special circumstances.

CHAPTER 2

Why Markets Are the Institution of Choice for Water Reallocation

"Water reallocation and marketing, for better or worse, are part of our future in the West, and we must prepare to face the challenge."

—Shupe, Weatherford, and Checchio 1989

Why, of all possible water-reallocation mechanisms, does this book focus on markets? From an institutional perspective, water can be allocated in any way permitted by existing aqueducts, pumps, and other infrastructure. Thus, a transfer of water is not necessarily a market-based transfer of water. A water transfer simply involves physical conveyance and perhaps altered use or altered timing of use of a flow of water. The water may be transferred by a single owner from one landholding to another, or it may be transferred to a new owner by means of inheritance, foreclosure, abandonment, government seizure and reallocation, land purchase, or even political consensus. There already exist numerous nonmarket mechanisms for allocating water, each with its own approach to implementation. Examples include the following:

1. *First-come, first-served allocation.* Queuing for water is the fundamental approach of the prior-appropriation doctrine. This doctrine was developed by eighteenth- and nineteenth-century immigrants to the

western United States who diverted river water for mining, agriculture, and other uses. It has the benefits of protecting historical uses of water and providing clear guidance under conditions ranging from drought to flood. It also permits relocation of use and changes in use. On the negative side, it protects uses and water-application methods that might have made sense a century ago but today are no longer appropriate. Indeed, the older the appropriative right, the more valuable it becomes.

2. *Allocation based on proximity to flows.* The location of one's land could determine water rights, as is the case in the riparian doctrine. Under the riparian doctrine, whoever owns land along a waterway has a right to draw water from the waterway. Essential (domestic) uses are given priority over irrigation and industrial uses during drought periods. Benefits of this approach include the fact that water is kept near the waterway, increasing the possibility that return flows will reenter the stream, and the fact that some connection exists between water available during dry years and the type of use (e.g., domestic use takes precedence during a drought). On the negative side, in areas where most water is found in large rivers separated by extensive dry lands (as in the arid West), development will be concentrated along the rivers, with no ready mechanism for moving water to more distant towns, farms, and businesses.

3. *Allocation based on deserving uses or users.* An intended use or some characteristic of the user could make an individual or enterprise "deserve" water. An example would be the reservation of water for fire suppression during a drought. One strength of this approach is that water is purposely allocated to important societal needs, perhaps during a drought emergency. Drawbacks include the challenge of establishing use priorities and keeping them up to date and the challenge of identifying acceptable ways to stop some existing uses in order to establish new uses.

Yet many argue that *market-based reallocation* is superior to all of these mechanisms. This chapter investigates that prospect. It begins with a general review of what is meant by a "market" and of the potential benefits of using a market. After a brief discussion of the many forms a market can take, the question of what a water market might look like is broached. The leading proposal for a large-scale water market, contained in California's Model Water Transfer Act, is presented.

The Market Institution

In the simplest terms, a market is a place in which goods and services are exchanged, either for other goods and services or for money. It consists of buyers and sellers. Of course, it is difficult to imagine a marketplace in which the flow of a river—surface water—can be purchased. For one thing, we do not typically think of the flow of a river as being owned and inventoried by sellers. For another, flows of rivers and other bodies of water are highly variable, changing according to precipitation patterns, cycles of snow and thaw, and other factors. In fact, these two characteristics make it particularly challenging to develop markets for surface water. But assuming for the moment that these challenges can be met, it is worth exploring both how markets function in a general sense and why they are such a common means by which human societies meet their material needs.

Markets perform a simple function: they allocate scarce resources among multiple users. They do so by setting prices. Prices force users to either pay up or make do without the resource, whether it is water, apples, or caviar. If the resource becomes more scarce—if, say, apples becomes less abundant because of a poor growing season—and insufficient in quantity to satisfy those willing to pay a given price, the price goes up until the resource clears the market. Only those who value the resource at the new price will purchase it. Those who would rather spend their money elsewhere will not purchase it. Likewise, if apples becomes more abundant, and more is available than is required at a given price, producers drop the price until those who would do without it at a higher price will go ahead and buy it. In this way, supply is matched with demand. In theory, at least, fruit does not rot in warehouses and hungry people do not starve (but they do not eat caviar, either).

Advocates for markets make one key claim for them: by providing crucial and timely information about the value of a resource (in the form of prices) to all producers and purchasers, markets enable members of a society to make decisions about producing more or less of a good, and about purchasing more or less of it, such that the society's resources are always put to their highest-valued use. In Adam Smith's classical economic theory, this is the famous effect of the "invisible hand" by which a society's resources are allocated in the most efficient way possible. Thus, the market is the means by which self-interested individuals allocate their resources in a way that promotes the common good.

This is the case in a theoretical, and for that reason perfect, market. In practice, markets are much more complicated, involving far more than buyers and sellers trading goods and services. A broader, more practical description of the market might be as follows: A market is *a set of institutions, expectations, and patterns of behavior that enables voluntary exchange.* Institutions can be defined as *the formal and informal rules that identify patterns, set boundaries, and establish consequences for human behavior.* Market-based transfers are based on the principles of willingness to pay and volunteerism, or choice. The most important (but of course not the only) element of a decision to participate in a market as a buyer or seller is price, which, as noted earlier, conveys a great deal of information on the relative availability of and demand for a resource. Price also provides comparative information so that market participants can choose among similar products. Relying on price signals has been shown in theory and practice to be an effective means to put goods and services into the hands of those who value them the most.

But some markets are extremely effective at allocating or reallocating goods (an example is the New York Stock Exchange), whereas others are not (early markets for industrial air pollution credits, for example, hardly ever saw a transaction). To understand the comparative effectiveness of different markets, one must focus on their institutional nature. As one might imagine, the sets of institutions required to implement the three approaches to water reallocation presented at the beginning of this chapter (as well as hybrid approaches) would be quite different from one another. Yet other institutions would be required to effect a market-based reallocation.

Perhaps the most important institution in a market-based reallocation is that of private property. Private property rights can be divided up in numerous ways. Table 2.1 provides eleven separate elements of a "full or liberal" property right as proposed by A. M. Honoré (1961), an innovative legal scholar. Only one attribute is essential to ownership. An individual may have all elements of ownership except the right to the capital (element 5) and still not fundamentally be the owner.[1] The right to the capital includes the power to alienate the property (to convey or transfer as a matter of an owner's choice) and the power to consume, waste, modify, or destroy it. The right to alienate includes the right to trade using a market. The right to consume, waste, modify, or destroy provides owners with additional flexibility not found in elements 1 through 4: to possess, use, manage, and derive income from the property.

A legal right to capital in combination with other elements of ownership provides existing and potential owners of water rights wide latitude to

TABLE 2.1
Elements of A. M. Honoré's Concept of Full or Liberal Ownership

1. *The right to possess.* The right to exclusive physical control of the thing owned.
2. *The right to use.* The right to personal enjoyment and use of the thing as distinct from (3) and (4) below.
3. *The right to manage.* The right to decide how and by whom a thing shall be used.
4. *The right to the income.* The right to the benefits derived from forgoing personal use of a thing and allowing others to use it.
5. *The right to the capital.* The right to alienate the thing and to consume, waste, modify, or destroy it. To alienate is to convey or transfer as a matter of an owner's choice and not in the course of the state's legal activities.
6. *The right to security.* Immunity from expropriation.
7. *The power of transmissibility.* The power to devise or bequeath the thing.
8. *The absence of term.* Indeterminate length of one's ownership right.
9. *The prohibition of harmful use.* Duty to forbear from using the thing in certain ways harmful to others.
10. *Liability to execution.* Liability to having the thing taken away for repayment of a debt.
11. *Residuary character.* The existence of rules governing the reversion of lapsed ownership rights. Reversion is the right of succession or future possession or enjoyment.

Source: Honoré 1961. See also Lawrence C. Becker, *Property Rights: Philosophic Foundations.* (London: Routledge & Kegan Paul, 1977), pp. 19–20.

generate value from their property. Someone who wants to sell water in a market must have strong enough property rights (though not necessarily full or absolute rights) to enable interested parties to anticipate a flow of value associated with a unit of property. It is this expected flow of value, in comparison with others, that forms the basis for exchange. In other words, the buyer must be assured that some other claim on the water will not interfere with the delivery of water he or she is purchasing. If a water right is to be of value to an urban area, it should include rights to consume, waste, modify, and destroy the water because this is often what happens to water that is subjected to municipal and industrial uses. A water right that excludes the right to capital takes on the quality of an instream-flow right, in which water use is permitted but not if it adversely affects the stream's existing flow regime. In urban settings, there is minimal demand for water with relatively truncated rights such as instream-flow rights.

Market institutions also give buyers and sellers some freedom in determining quantity, quality, and terms of exchange—*some* freedom, not *com-*

plete freedom. Not only do laws and social customs place limits; inherent characteristics of the product exchanged, or complementary products, may also place limits on terms, quantity, or quality. In the case of surface waters, terms of trade must take into account the variability in annual supply. It is likely that no large-scale water-transfer contract would *under any conditions* guarantee delivery of a specific quantity of water. Similarly, some uses, such as urban provision of potable (drinkable) water, might require a higher-quality delivery, whereas other uses, such as irrigation of orchards, could be satisfied with a slightly lower-quality delivery. In the former case, terms of trade could place limits on sources or prior uses of delivered water in order to protect the quality of the water.

Laws and social customs do not merely limit exchange: they also enable it. Governments recognize and enforce property rights; provide recourse in cases of disputes, malfeasance, or damages; establish authoritative weights and measures; provide and regulate currencies; manage transportation, communication, and other infrastructure; maintain public order; and provide other services critical to a market system. In the case of water, the governmental role is indispensable in managing multifunction dam-reservoir systems (i.e., balancing flood control, environmental protection, recreation interests, and drought protection). The government also sets and monitors limits on industrial effluents and other pollutants that could harm the resource. That is, today's popular dualism of "markets versus government" is misguided: without extensive institutional support that can be provided only by government, few modern markets for anything would exist.

The nature of the good itself also provides constraints and creates opportunities. for exchange. There is a well-known category of goods called *public goods,* the benefits of which are spread among many parties and cannot be denied to any party. Street lighting and national defense are the most common examples. Individuals typically do not purchase and install street lights because they cannot exclusively utilize the benefits of doing so. Instead, collective action is taken to provide these and other public goods, with general taxation spreading the cost among all beneficiaries. Water that is used to provide environmental benefits, such as species preservation, is just such a public good. It is possible that unless governments themselves participate in water markets, a market might underprovide public-goods-type benefits of water.

If a good is to be tradable, it should be divisible, with identifiable boundaries that can be used to apportion it among different owners. Its dimensions should be measurable, not just in a laboratory but also in contexts that allow its transfer to proceed. Further, it should be separable from its existing physical (in the case of water, biogeophysical) surroundings without loss of value. Given water's liquid nature, its annual variability of supply, its ability to be reused repeatedly, its ecological importance to its conveyance system, and the massive quantities involved in long-term transfers (much larger in volume than the world's largest skyscrapers), it presents unique physical challenges as a market-available commodity, but these challenges are not insurmountable. Although no other commodity is exactly comparable to water, similar cases exist: there are, for example, efficient, large-scale international markets for oil, as well as much-improved markets for rights to emit air pollutants. The existence of a small number of actual long-term water transfers in California (and more elsewhere) suggests that practical methods of dealing with the physical issues of market-based water transfers can be identified.

Arguments in Favor of Market-Based Water Reallocation

When it comes to justifying the choice of a market mechanism, the institution of private property, with its own social implications, is tightly intertwined with market-based reallocation. Three different qualities characterize a well-functioning market based on private property suitable for redistributing water: economic efficiency, enhanced individual freedom, and availability of additional policy options.

Economic Efficiency

A primary benefit of market-based reallocation is that it promotes economic efficiency, a goal strongly emphasized in economic theory. The strongest form of economic efficiency is Pareto optimality, in which no further reallocation of goods can be made that would benefit one party without hurting another party. That is, in a Pareto optimal economy, with a given starting point of who owns what, parties trade in such a way as to meet their preferences as well as possible, but no one makes a trade (or is forced to make a trade) that leaves him or her less well off. All traders are then as well off as they can be. A slightly weaker version of economic efficiency, called Kaldor-Hicks efficiency, is achieved when all possible trades have been concluded that involve net benefits to society. As a whole, winners benefit more than losers suffer, so society is better off. Winners must, however, directly compensate losers for their losses. Markets can accomplish efficient reallocation at a cost lower than those of alternative mechanisms because informed volunteerism is built into the system. The government is not put in the awkward position of trying to figure out what people's preferences are and then allocating resources to satisfy those preferences. In a perfect market, people make their own decisions and no one is forced into an exchange that he or she does not see as beneficial. Further, as prices (and preferences) change, buyers and sellers have the flexibility to return to markets and readjust their "portfolio of assets."

Enhanced Individual Freedom

Law professors Robert Cooter and Thomas Ulen (1996, p. 109) summarize the relationship between markets and liberty succinctly:

> Control over economic life provides [governmental] officials with leverage that can be used to control other aspects of life, whereas private property creates a zone of discretion within which individuals are not accountable to government officials. Private property [the necessary precursor to markets] has thus been viewed . . . as a bulwark against the dictatorial authority of governments.

Arguments similar to this are found in the Federalist Papers, published in 1787–1788 to build support for ratification of the Constitution of the United States.

With respect to water, a theoretical benefit of a market could include a decision by environmentalists to purchase water rights in order to preserve a valued wetland or increase a waterway's springtime flow.[2] Without a market mechanism, environmentalists would have to rely on governmental mechanisms to achieve the same end. Thus, the "zone of discretion" created by market mechanisms can be used to achieve socially valued ends.

Additional Policy Options

Negotiations in California and elsewhere—not only negotiations about water but also those about restructuring of the electricity sector and other institutional reforms—often involve working groups: representatives of interested parties who meet regularly to bring specialized expertise and contrasting interests to the table and then prepare reports and position papers for regulatory bodies. The regulatory and oversight bodies, such as the California Public Utilities Commission and the California Department of Water Resources, then either take regulatory action or forward recommendations to the state legislature. Working groups provide a venue for adversarial parties to identify and define issues and to attempt to reach consensus (some might say to "make a deal") on proposed reforms.

At the basis of this process is an expectation that a general meeting of the minds will eventually occur between adversarial interests or, failing that, that the losing side will at least be able to carve out some protections or other benefits from an otherwise unfavorable policy. In short, a policy should emerge with which the losing side can live. The working-group process is necessarily a slow one because even experts need time to grasp complex emerging issues; to realize, in many cases, that they are on the losing side of the policy issue; to reformulate their policy strategies; and to move toward consensus.

Occasionally, differences in interests are so clearly drawn that policy-making gridlock results. No solution meets minimum requirements, distrust skyrockets, and collegiality among adversaries breaks down. Although legislative and regulatory bodies certainly are empowered to act under these circumstances, their members prefer not to, anticipating lawsuits and political fallout if they do. When gridlock occurs, market-based solutions may offer new ways of thinking that increase the number of potential solutions and lead to a compromise. More generally, in places where market mechanisms have social legitimacy (as in the United States) and regulators have the necessary combination of experience, expertise, and flexibility, market-based regulations may prove to be feasible and appropriate when traditional regulations do not provide solutions.

Needless to say, water-allocation issues have many characteristics that invite controversy: multiple demands on a scarce resource, few overlapping interests to form the basis for "horse-trading," conflicting visions of optimal future paths for one's state (e.g., urban growth versus rural protection versus environmental conservation), and histories of adversarial relations and distrust among interest groups. Market-based systems, by increasing the number of possible outcomes, could therefore figure prominently in resolutions of current water-allocation controversies.

The Many Forms That Markets Take

Markets take countless forms around the world. Some markets exist only among huge, well-financed firms (e.g., markets for airport landing slots), and some markets exist *within* huge firms (e.g., internal labor markets). Some involve transactions between individuals (flea markets). Some markets are structured like auctions, whereas others involve extensive private negotiations between buyers and sellers (markets for computing services). Some transactions require a great deal of paperwork, and others need only a handshake. Some transactions require extensive follow-up and monitoring to ensure that each side sticks to the terms of the agreement (quality control of numerous products), and others are executed without any oversight at all. Moreover, some companies transact over and over, year after year, with the same partners, whereas other companies transact only once with each other and have no subsequent contact.

What accounts for the wide variety of market institutions that exist from product to product and from region to region? As mentioned earlier, markets are one type of institutional solution to the practical problem of allocating scarce resources. An explanation for the various forms taken by markets and firms was suggested in 1937 by economist Ronald Coase: parties will choose the set of institutions that minimizes their cost of transacting (Coase 1937). For example, if an automobile manufacturer fears that a supplier might provide substandard parts in order to cut costs and reap more profits, the manufacturer might choose to merge with the supplier. Then both sides have an interest in maximizing the profits of the co-owned activity. The combined entity spends less money monitoring performance, and more money is available for achieving a successful commercial outcome.

A large array of factors influence the cost involved when two parties transact. In some cases, transaction costs are relatively low. If neighboring farmers agree to transfer a water allotment for one season, the agreement consists of little more than a telephone call between them followed by a note to the irrigation district, which handles the accounting and timing of delivery. The combination of shared water infrastructure and a within-district transfer allows them to transact at little cost, so there is no reason to merge their operations in order to trade water. Alternatively, an irrigation district might attempt to transfer water rights to an urban water agency on a permanent basis. The city might be expanding and searching for new long-term sources of water. In this case, with the prospects of permanent loss of water in an agricultural region and altered environmental conditions due to a redirection of surface flows, the level of regulatory and social scrutiny would be significantly higher. It might be necessary for the two public agencies to merge their activities by creating a joint powers authority (JPA), or merged governmental authority. JPAs can create a context in which transactions are more likely to be completed.

This book focuses on the long-term reallocation of water rights and water use from current uses and locations to alternative future uses and locations. Agreements between buyers and sellers of water consistent with long-term reallocation could take many forms, some more and some less desirable from a societal perspective. One can envision the purchase or long-term lease of appropriative rights or contract rights or the purchase and retirement of agricultural land, with water rights either being transferred away, being used to support urban growth on the same land, or being applied to environmental restoration of the same land. Further possibilities include the creation of long-term option contracts in which water use is shifted from farms to cities during dry years. The parties to these transactions are likely to include water agencies or state or federal environmental authorities as buyers and some combination of individual farmers and their irrigation districts as sellers.

Environmental organizations, however, are not likely to be major participants in long-term water-reallocation transactions. Environmental organizations cannot tax residents to raise money for water purchases, so their pockets are likely to be less deep than those of water districts. Moreover, the cost of building and coordinating a "buyer's coalition" of environmental groups can be prohibitive. Further, because water serving in an environmentally supportive capacity is a public good, there is a high risk of "free riding," in which some potential beneficiaries count on others to pay the cost of purchasing water. Water purchases for environmental purposes are therefore likely to be carried out by state or federal governmental agencies, where sufficient financial means exist and the cost is spread among a far wider range of beneficiaries.

The Leading Market-Based Alternative: The Model Water Transfer Act

The leading proposal for a water market in California is embodied in the proposed Model Water Transfer Act for California. Although as a state senate bill in 1997 (S.B. 15) its life in the legislature was relatively short, the Model Act provides a detailed framework for institutional reform. It remains influential because it still represents "best thinking" on the part of the state's most established pro-market interest groups and academic thinkers. The following is a brief review of its provisions related to long-term water transfers.

In its initial section, the Model Act embraces voluntary water transfers because they provide flexibility in water reallocation and because they promote reallocation based on the principle of economic efficiency. The importance of protecting other parties who might be adversely affected by water transfers is also stated. These parties include other holders of water rights, other lawful water users, local water agencies, cities and counties, and other persons who use the state's water resources for their livelihood, recreation, and aesthetic enjoyment. There is a second list of potentially injured parties and uses: water users, water quality, fish and wildlife, other instream uses of water, groundwater resources, and regional economies. All of these interests are described as "third parties."

The operating assumption of the Model Act is that administrative processes that now govern long-term marketlike water transfers are too cumbersome and inclusive, thus limiting essentially to zero the number of proposed transfers that are satisfactory to enough stakeholders that they can proceed. If the administrative process of transfer review and approval can be streamlined, more beneficial transfers will occur.

Under the Model Act, review and approval would be expedited primarily by restructuring and limiting the input of potentially injured parties. Water-transfer negotiations would be exempted from certain provisions of the California Environmental Quality Act, and there would be a new process for comment and review. One key element of the process would require any party wishing to comment officially on a proposed transfer to have filed an initial protest soon after submission of the transfer proposal to the State Water Resources Control Board. Parties who subsequently realize that they might be injured by a proposed transfer would not be provided access to the review process. Other avenues of challenge to proposed water transfers, such as lawsuits, would also be limited, as would the definitions of who has legal standing to challenge and how much in damages can be collected.

With new limitations placed on outsiders' participation in the transfer-approval process, the Model Act anticipates both that more transfers will occur and that some of them will be damaging to other parties. Under the act, injured parties could seek payments from a Water Transfer Compensation and Restoration Fund. The efficiency envisioned in the Model Act is therefore of the Kaldor-Hicks variety, whereby winners directly compensate losers and society is better off overall. However, since a cap is placed on the total liability of parties to the transfer (no more than $30 per acre-foot transferred), full compensation may not be available if numerous claims are recognized. Claimants would have to demonstrate that their injuries were caused by the water transfer itself and not by other factors, such as drought or other water shortages. The claims process could occur years after transfer negotiations are completed if physical delivery of water is postponed by the buyers. If the transfer agreement includes annual transfers of water, injured parties' claims on the fund would have to be made on an annual basis.

Another key set of provisions involves water as property. The Model Act would strengthen ownership rights to water. For example, after long-term but not permanent transfers, the original owners would have full rights to recover and resell their water rights, even if the water had been made available through conservation measures and could no longer be applied beneficially to its original purpose. The goal is to increase the confidence of rights owners that if they choose to part with water rights temporarily, their doing so will not be taken as a demonstration that they do not really need the water and therefore should eventually give it up.

The major aims of the Model Act in relation to long-term transfers, then, are to streamline the process of transfer review and approval by limiting the ability of third parties to challenge proposed transfers and to concentrate water rights in the hands of legal owners of water. When transfers do injure third parties, claims can be made on a compensation fund. The changes proposed in the act are based on the belief that the resulting enhanced economic efficiency of market-based water transfers will be great enough to compensate injured parties, resulting in a net gain to society.

California's Model Water Transfer Act is the most comprehensive how-to statement with respect to creating water markets. Other proposals have dealt with such issues as how large the boundaries of the market should be, ranging from relatively small (e.g., only in a watershed or only among current contractors for State Water Project deliveries) to huge (e.g., including all of western North America). None of these other proposals have engaged the details of market creation to the extent found in the Model Act.

Conclusions

Decisions must be made at many steps in the societal process of selecting new users of scarce water resources. The first step, understanding whether water will be reallocated at all and, if so, from whom and to whom, was the subject of chapter 1. This chapter examined the second step: deciding what *kind* of market should be established. It defined the terms *market* and *institution,* and explored some of the common justifications for use of markets. It described how and why markets take the many forms they do and then presented some of the logic and mechanics of the leading proposal for a market for long-term transfers of water. So far, the leading approach, which has been around in general form since the 1970s, has not succeeded in bringing about a market-based reallocation of California's water. Understanding why the leading market-based approach has been unsuccessful so far is an important next step before focusing on the challenge of constructing a new water-reallocation proposal.

Notes

1. See Lawrence C. Becker, *Property Rights: Philosophic Foundations* (London: Routledge & Kagan Paul, 1977), pg. 19–20.
2. It is a *theoretical* benefit because, as described elsewhere in this chapter, this behavior runs into "public goods" problems that could limit the environmentalists' ability to raise enough money to purchase sufficient rights.

CHAPTER 3

Political, Legal, and Cultural Challenges to Water Markets

"There is a lot more to water marketing than meets the eye."
—Porgens and Carter 1992

As described in chapter 2, effective markets do a number of things well. They facilitate the allocation of goods to those who value them most in terms of willingness to pay. A resource-allocation system that helps produce an economically efficient outcome (i.e., a well-functioning market) requires fewer resources to achieve the same level of use benefits than do other allocation systems. This characteristic is especially valuable when a *natural* resource is being consumed. In general, the less water that is pulled from natural systems (rivers, wetlands) into human-made systems (canals, reservoirs), the better is the health of the natural systems. Therefore, any allocation system that makes the most of every drop of water is indirectly serving the ecological interests of the natural supply system. Further, markets establish zones of noninterference by governmental entities, and they provide additional policy options when society is faced with a challenging resource-allocation issue. These important benefits help explain why policy makers and many interest groups advocate markets for the reallocation of freshwater and other scarce resources.

But markets also have their limitations. Some of the limitations are due to the fallibility of the humans who devise and administer markets, and others are due to inherent and unavoidable qualities of the mechanism. The inherent problems of markets in general, and water markets in particular, have been well known for years. For example, markets distinguish between individuals who are buyers and sellers and those who are not. Those who are not directly involved in a transaction as buyer or seller may be affected by it, but they have little or no say in whether the transaction proceeds. Known in economics as third parties, they experience effects that are called externalities. Farm-sector externalities resulting from a major rural-to-urban water transfer might include loss of farming jobs, reduction in the number of farmers paying to keep up a shared irrigation system, unemployment-related burdens on county or city governments, and a drawdown of shared groundwater resources if a seller of surface-water rights switches to groundwater use. Because markets operate on the premise that duly empowered sellers and buyers will transact between themselves, third parties and externalities are bound to emerge in transactions involving a multiple-use natural resource such as water.

Other inherent problems exist. Market mechanisms do not automatically protect specific sectors of the economy or society or specific geographic regions. As the old saying goes, water flows uphill (or anywhere else) to money. Since the rich are willing to pay more than the poor, the rich will probably own more water. Markets also do not protect uses of a resource that promote general rather than individual welfare (the public-goods problem introduced in chapter 2). For example, large numbers of citizens might be willing to pay a share of the cost to restore water to a given scenic river, but most likely no one citizen would be willing to pay enough for such a transaction to occur. Organizing for collective action on the part of buyers is time-consuming and costly; indeed, few individuals have an incentive to undertake the task at all.

With respect to natural systems, there is a critical variable that could turn potential environmental benefits associated with efficiency (described earlier) into potential environmental disasters. It is the rate of growth in use of the resource. Is the quantity of an economy's resource use, or resource throughput, stationary or growing? If it is stationary (no net growth in resource use) and the issue is one of selecting a resource-allocation mechanism to utilize a fixed amount of available resource, then an efficient system (such as a market) probably would be neutral or perhaps would even benefit the ecosystem that supplies the resource. But if the level of resource use is not fixed but instead is growing as a region's population, affluence, or economic expectations grow, then an efficient system could simply accelerate the consumption of a natural resource well beyond its sustainable yield. That is, assuming that no regulations limit the exploitation of natural resources, an efficient system will identify opportunities for profitable exchange involving natural resources more quickly than will an inefficient system, thereby accelerating the throughput of natural resources through the economy.

Thus, a market mechanism, though demonstrating many positive attributes with respect to resource allocation, also has inherent limitations. Through careful institutional design, it is possible to avoid or minimize the negative effects of a market mechanism, though whether the negatives are minimized enough to warrant the use of a market is a matter for the public to decide. The balance of this chapter explores three attributes of markets as applied to water reallocation that, if ignored by policy designers, could result in unrealistic and ineffective proposals for institutional reform: (1) the political nature of the process of market creation; (2) the importance of aligning market reforms with historical trends; and (3) the power of markets to adapt quickly and without conscious human oversight. These aspects of the market mechanism, combined with the inherent limitations listed earlier, must be accounted for if efforts to create a water market are to be effective.

The Political Nature of Water Markets

If a water market does emerge, it will be because political bodies have created new institutions to enable a price-based, voluntary reallocation of water. When public and private interests (urban water agencies, environmental interests, industrial interests, farming interests, etc.) get involved in the process of market creation—through legislation and regulation and even through the courts—there is certain to be political and economic self-interest behind the recommended forms of institutional change. These interests are not out to design a Pareto optimal market mechanism. Rather, they have a stake in particular allocation outcomes, and they will propose institutions that further their goals.

There is likely to be a predetermined directional nature to the flow of water that will be exchanged in California water markets. Urban, industrial, and environmental interests located primarily in the Greater Los Angeles area and the San Francisco Bay area are deeply concerned with water reallocation. The "market outcome" most important to them is precisely the movement of additional water resources to urban areas and environmental uses. Alternatively, agricultural interests have their own directional preferences for the way water is reallocated: within and not beyond agricultural regions.[1] As described in appendix 1, the guiding state policy favors the urban, industrial, and environmental interests. The state is seeking ways to move water out of the agricultural sector and toward growing urban populations and environmental uses.

To be sure, a directionally predetermined market mechanism by design will not capture all potential cost savings from economically efficient trading. Being a market mechanism, however, it will capture some cost savings: those that are available in the context of the overarching reallocation policy. Aside from the conscious choice to forgo some potential gains from trade, there are no inherent problems with using a market mechanism to carry out water reallocation according to a predetermined plan, and there may be numerous benefits. If a valid political process has yielded the larger policy goal, efficient policies in pursuit of that goal should be welcomed. In his book *Beyond Growth,* economist Herman Daly (1996, pp. 52–53) favors market-based reallocation of natural resources, provided there is a cap on the overall throughput of the resource:

> [A]fter having made social decisions regarding an ecologically sustainable scale and an ethically just distribution, are we in a position to allow reallocation among individuals through markets in the interests of efficiency.

Daly's concern here with respect to scale, or quantity, of resource throughput is that it should not be left undetermined; rather, it should be chosen in accordance with sustainability criteria. When this is done, markets are utilized to enhance efficiency of use.

Indeed, if policy makers seek to create ideal, efficient market institutions that capture all possible gains from trade, they may be frustrated by the political roadblocks to reform. The major interests behind the push for water reallocation may oppose an efficient market because it does not adequately guarantee the outcome they desire. Opponents to water reallocation probably also would oppose a market designed to achieve maximum efficiency. Opposition from essentially all interested parties is enough to kill any proposal. A water market designed with clear reallocation goals in mind is more likely to be consistent with state policy, as well as to satisfy the major interested parties, than is a theoretically more efficient market whose outcome cannot be predicted.

There is also a regional political context to water markets. Because water is important not just to the primary users but also to the region where the use is occurring, any set of institutions that enables the reallocation of water will have political effects on water-losing and water-gaining regions. Water-losing regions in California, typically located in the northern and eastern portions of the state, will encounter more difficult economic-development challenges in the future than will the water-gaining regions (coastal urban areas). Because the extent of a region's economic power is one element of the extent of its political power, a policy that redistributes future economic opportunity certainly is political in nature. The potential political outcomes of the choice of a market mechanism should be considered alongside other criteria when evaluating markets.

Aligning Market Institutions with Evolving Property Rights

As described in chapter 2, effective market mechanisms require that property rights be concentrated in an empowered decision maker (seller). The set of rights attached to a particular pattern of water use should be comprehensive enough that an outsider could reasonably anticipate a future flow of value from ownership of those rights. The transaction then becomes an exchange of wealth today (a payment) for rights to a potential flow of future wealth defined by the set of property rights. In certain contexts, this fundamental quality of markets can lead to valid objections to their use.

The process of concentrating transferable property rights in a single owner is also the process of creating third parties and situations that may involve externalities. If an individual has a say over a transaction—over its structure, timing, magnitude, or other factors—that individual is not a third party, and the consequences of the transaction are not externalities. Even if the say is not definitive, as long as the affected individual has some (nontrivial) influence, he or she is not excluded from the transaction process.

An issue of fairness arises in the question of whether and how to concentrate property rights: when market institutions are crafted in a conscious manner (through a public-policy process), the case must be made that social benefits arising from concentration of the property rights (as a precursor to market creation) exceed the negative effects on individuals and groups that have lost their portions of the formerly diffuse water rights. Environmental, rural-town, family-owned farming, recreational, and other interests could be negatively affected, for example, if tradable water rights were concentrated in the hands of a region's larger farmers or the irrigation district and then sold.

Just as concentration of property rights creates third-party and externality issues, diffusion of rights reduces these effects because third parties are given partial ownership, which affords them some control. *Diffusion* here means taking absolute control over property rights away from a single owner and allowing other parties to have claims on the good—on the way it is used, whether and how ownership can be transferred, and so forth. The more completely water rights are distributed, or withheld from the primary owner, the more those rights can be described as diffused. Typically, pro-market academics, advocates, and policy makers describe diffused rights as "poorly defined." What they really mean is "not defined in a manner that permits market-based transfers." A fuller meaning of diffused property rights is that the interests of a wide range of individuals and parties are protected by their having some measure (albeit small) of property rights to a crucial natural-resource system such as water.

This idea that the diffusion of property rights can protect the public interest is consistent with the way California's judicial system interpreted property rights in water over the course of the twentieth century. Clifford Schulz and Gregory Weber (1988, p. 1065) summarize the changes in California water law:

> [T]he courts are propelling California into a new era of judicially and administratively supervised reallocations of its water resources, on the premise that water use is more a governmentally granted privilege than a privately held property right.

What does this new era mean in terms of use of a market mechanism to reallocate water? The following is an examination of the consistency of market-based reallocation of water resources with the historical evolution of water rights in California.

Erosion of Initially Strong Individual Rights

Water law expert Joseph Sax (1990) describes water law as a story of evolution to meet the changing needs of society. During the nineteenth century, California's public-policy needs called for the development and diversion of water. In an era of competition among western states for scarce capital and entrepreneurial talent, legal institutions supportive of business were established (Bakken 1983). In 1859, the Supreme Court of California recognized that ownership of water, "as a substantive and valuable property right" that "may be transferred like other property."[2] By 1912, the court had added, "[T]he water-right which a person gains by diversion from a stream for a beneficial use is a private right, a right subject to ownership and disposition by him, as in the case of other private property."[3] Clearly, the court conceived of water rights as voluntarily transferable, most likely through market-like institutions.

Although afforded strong protection from regulatory intervention, water was not completely like other property, even in the nineteenth century. Then, as now, appropriators could change the place, time, or purpose of diversion or return flow, but only in ways that would not injure other appropriators. The "no-harm" (or "no-injury") rule essentially gives an appropriator a vested right to the stream's flow characteristics at the time of the appropriation. Although exceptions exist (for water imported into one watershed from another and for water released with the prior intention of recapture and reuse), the no-harm rule still serves to limit interbasin transferability of certain rights, depending on the proposed transfer's specific circumstances.

By 1914, the state had involved itself in the registration of rights, opening private water use to public scrutiny for the first time.[4] Both registration of rights and, by the 1950s, a permit-application process for new appropriations served to increase the separation of water rights from other forms of property rights.

Reasonable and Beneficial Use Requirement

At least two doctrinal shifts since the 1920s have directly influenced the tradability of water. They are the reasonable and beneficial use requirement and the public trust doctrine. In the mid-1920s, Southern California Edison, one of the state's largest electric utilities, proposed damming the San Joaquin River to produce hydropower. A riparian water user sued to block this plan because she needed the flow of the entire spring flood to irrigate her pastureland. In *Herminghaus v. Southern Cal. Edison,* the California Supreme Court agreed with the riparian user and blocked construction of the dam.[5] What followed was public outrage that an insignificant riparian user could block the further electrification of Greater Los Angeles, culminating in a 1928 amendment to the state constitution. The amendment, now located in Article X, Section 2, requires that all water uses be reasonable and beneficial and that uses of water be evaluated for wastefulness. Although the *Herminghaus* case did not involve a water transfer, Schulz and

Weber (1988) believe that the pro-property decision firmly established in California law the precedent that water rights were compensable property interests with well-defined scope and limitations. The subsequent constitutional amendment cut deeply into water-rights holders' right to their capital: water owners' rights to consume, waste, modify, or destroy their water were now limited well beyond the prior no-harm limits. Other property rights curtailed by Article X, Section 2 include the rights to possess and manage water resources and the right to security from expropriation by the state.

In 1967, some forty years after passage of Article X, Section 2, the California Supreme Court further weakened private property rights to water. The case of *Joslin v. Marin Municipal Water District* involved a sand and gravel company located alongside a river (and therefore holding riparian water rights) that depended on the river to carry the sand and gravel to its property. There, the sand and gravel was excavated and sold. When a water district sought to build a dam upstream that would impound the sand and gravel, the company sued. The judge decided on behalf of the water district, basing the decision on the belief that the sand and gravel company's use of the stream was unreasonable.[6] As Schulz and Weber (1988, p. 1089) note, "*Joslin* seems to hold that the issue of which competing water user possesses the superior right is no longer to be decided by reference to riparian rights or temporal priority, but by which water use best serves perceived public policies at the time the dispute is resolved." In other words, current public interests in water use were deemed more important than established rights when disputes arise. This amounted to a diffusion of property rights to water.

This theme was reiterated and extended two decades later in the Racanelli decision, related to water use affecting California's San Francisco Bay/Sacramento–San Joaquin Delta (Bay-Delta) system. Here, Judge John Racanelli cited both Article X, Section 2, of the state constitution and the *Joslin* case in support of extending the power of the State Water Resources Control Board (SWRCB) to cancel or reorder the priority of appropriative rights for purposes of improving water quality. How the changes were to occur, he wrote, was to be "a policy judgement requiring a balancing of the competing public interests."[7] His opinion provides a public-benefit alternative to riparian and appropriative rights doctrines for sorting out competing claims to water. Schulz and Weber consider the Racanelli decision to be "the current high-watermark of the post-1928 . . . assault . . . upon vested property rights in California resources" (1988, p. 1093).

The expansion since 1928 of the reasonable and beneficial use requirement imposes important limitations on the potential effectiveness of a water market. If current owners of water rights do not have secure rights—even if the lack of security serves perfectly valid public purposes—they will have a difficult time finding buyers for those rights. At a minimum, a buyer would lower considerably the amount he or she would pay because of the reduced security of ownership. But given the central importance to urban areas of secure supplies of water, a significant possibility that water rights will be lost is likely to result in no transaction at all.

Public Trust Doctrine

The second doctrinal shift can be traced to the 1983 California Supreme Court decision in *National Audubon Society v. Superior Court of Alpine County*.[8] In this decision, the public trust doctrine, which requires states to serve as trustees of the navigable waters and underlying beds within their states, was integrated into the appropriative-rights system. Attorney Roderick Walston (1989) outlines the four principles that define the expanded public trust doctrine: (1) the state as sovereign retains continuing supervisory control over navigable waters and underlying beds; (2) the legislature has the right to grant usufructuary (use-only) water rights even though such rights will not promote, and may unavoidably harm, the trust uses at the source stream; (3) the state has the affirmative duty to take the public trust into account in planning and allocating water resources; and (4) the state has a duty of continuing supervision over water rights, even after such rights have been granted. According to Gregory Weber (1994, p. 923), the case asserted the state's "manifest and continuing public control over water allocation," including promotion of navigation, hunting, fishing, commerce, aesthetics, recreation, and preservation of ecosystems in their natural state. As of 1994, eight different sections of the California Water Code defined aspects of public uses of water under public trust, and courts retained the right to make decisions on public-trust grounds. Thus, property rights to California water had again been diffused. Far from nineteenth-century "vested property rights," water users' rights now were largely usufructuary, attenuated by public-trust requirements, and subject to continuous public oversight. To state the obvious, such rights are hardly consistent with transfer from one owner to another through a market.

Area-of-Origin Protections

Another diffusion of property rights occurred in the 1930s with the passage of area-of-origin protections. Partly in reaction to the economic weakening of the Owens Valley following its loss of water rights to the city of Los Angeles, a 1931 statute provided counties from which water is exported a "right of recapture" of the water at any time if necessary for the development of the county.[9] Similarly, a 1933 statute gave the watersheds and adjoining areas from which Central Valley Project (CVP) waters originate a standing prior right to recall "all of the water reasonably required to adequately supply the beneficial needs of the watershed, area, or any of the inhabitants or property owners therein."[10] Given the centrality of water to economic development in California, these laws attempted to protect the economic future of northern and eastern counties, given the historical accident that the southern and western parts of the state had been the first to develop.

Today, area-of-origin protections mean that purchasers of otherwise valid long-standing water rights could find themselves in costly and contentious legal confrontations. A key issue involves quantifying true area-of-origin development-related water needs. Also unresolved are the questions of which areas qualify as areas of origin, whether protections can be applied in equal measure against both state and federal existing uses, and what, specifically, are the protections provided.

The 1959 California Water Resources Development Bond Act, also known as the Burns–Porter Act, attempts to lay groundwork for easing adverse effects when rights are eventually recalled by areas of origin. In addition to authorizing funds for initial construction of the State Water Project (SWP), Burns–Porter includes a plan to set aside royalties from offshore oil and gas production for future construction of northern California water-supply projects.[11] The future water developments are meant to replace water recalled under area-of-origin statutes. Ronald Robie, then the director of the California Department of Water Resources, and Russel Kletzing (1979) pointed out, however, that by 1979, the royalty revenues were only a fraction of what would be needed when northern parts of the state began exercising their rights. From the perspective of a potential buyer of long-term interregional water rights, then, area-of-origin protections diminish the value of the property by creating insecurity of ownership and the potential for future litigation should the water be recalled.

Groundwater Doctrine

Property rights to groundwater have similarly been diffused over time. The starting point for groundwater doctrine is the 1843 decision in *Acton v. Blundell,* an expansive grant of private-property rights by the English Court of Exchequer Chamber.[12] In *Acton,* a landowner was held to own everything below the surface of his land and could therefore pump and use groundwater regardless of the effect on neighbors. California upheld *Acton* in some late-nineteenth-century cases but rejected it in the 1903 *Katz v. Walkinshaw* decision, which called for "giving to each [user of a shared aquifer] a fair and just proportion" during periods of shortage.[13] This decision hinged in part on recognition that the hydrogeology of California had different legal requirements from that of England (Grover and Mann 1991). In England, for the most part, rivers run year-round and rain falls year-round. In California, many rivers dry up by late summer and rains do not return until late autumn. Moreover, California's rivers, though large, are more dispersed than England's. This increases the dependence of many Californians on aquifers as a sole source of water for part of the year, as opposed to the aquifers serving as a year-round supplemental source. Sharing the yield of aquifers during a drought, regardless of prior rights, is therefore more defensible in an arid region.

From the 1950s through the 1970s, California appeals court rulings further eroded rights of prior appropriators of groundwater in favor of administrative-sharing arrangements. In 1956, an appeals court validated a district's combined plan for water importation and aquifer recharge and its level pricing policy for all pumped water.[14] Since all groundwater users paid the same price to maintain water levels in the aquifer, the benefits of earlier appropriation of groundwater rights were lost. In a 1974 case, a private company pumping groundwater lost its pumping right when it interfered with a water district's groundwater-recharge program.[15] One year later, the court apportioned an overdrafted groundwater basin, basing its decision in part on the reasoning followed in a federal case, *Nebraska v. Wyoming,* in which the judge took into account equity factors such as physical and climatic conditions, consumptive uses in different areas, and the extent of established uses (Garner et al. 1994).[16]

Each of the doctrines and protections described here tells a tale of the diffusion of property rights to water over the course of the twentieth century. Groundwater experts George Grover and John Mann Jr. (1991) believe that in the case of groundwater courts have recognized that changes in factual circumstances, in legal concepts, and in economic and social conditions call for equally dynamic changes in law. Table 3.1 lists many of the effects of legislative and judicial actions on water rights in California, presented within the framework of A. M. Honoré's elements of full or liberal ownership (see table 2.1). For example, absence of term (element 8) is limited by the reasonable and beneficial use requirement, which could result in forfeiture of a water right. Other rights, including the right to cap-

TABLE 3.1

Some Abridgments to Honoré's Elements of Full or Liberal Ownership as They Relate to Water in California

1. *The right to possess.*
   - Original purpose, point of diversion, and return flow limitations
   - Federal navigational servitude
   - Environmental restrictions to protect instream flows and water quality
   - State public trust limitations
   - Riparian possession limited by location of landownership

2. *The right to use.*
   - Surface rights are shared
   - Limited to original purpose
   - Districts possess water rights on behalf of landowners and farmers
   - Public trust limitations
   - Environmental restrictions if use adversely affects water quality
   - Junior appropriators lose use rights in dry years
   - Riparian users limited to domestic use in dry years
   - Riparian use rights limited to property bordering the waterway and in the watershed

3. *The right to manage.*
   - Appropriators limited to original purpose of the appropriation
   - Reasonable and beneficial use requirement

4. *The right to the income.*
   - Appropriative doctrine forbids profit-motivated speculation
   - Easiest when water and land are leased together

5. *The right to the capital.*
   - Alienation (i.e., transfers) limited by no-injury rule, environmental regulations, and informal community claims
   - Consumption limited by terms of the permit
   - Waste forbidden by reasonable and beneficial use requirement
   - Modification, such as increased salinity in farm runoff, is subject to concentration limits
   - Destruction is allowed only if written into the permit

6. *The right to security.*
   - Public trust doctrine reduces security
   - Area-of-origin protections reduce security
   - Federal navigational servitude reduces security
   - Requirement for reasonable and beneficial use reduces security

7. *The power of transmissibility.*
   (no specific abridgments)

8. *The absence of term.*
   - Rights can be forfeited or abandoned on the basis of patterns of use

9. *The prohibition of harmful use.*
   - No-injury rule and environmental regulations consistent with this prohibition

10. *Liability to execution.*
    - Most liable when attached to land

11. *Residuary character.*
    - (no specific abridgments)

Source: Elements of ownership from Honoré 1961.

ital, are subject to numerous abridgments. Among them are the no-harm rule, the requirement to abide by one's permit or original use claim, and the requirement to use water in a reasonable and beneficial manner.

If primary owners of water rights (farmers, irrigation districts, state and federal water projects) no longer possess nearly complete ownership of their water, who does? Two things could have happened to rights no longer assigned or only partially assigned to primary owners. First, rights could have been redistributed to others. For example, the no-harm rule extends weak capital rights (in the form, for example, of legal standing to pursue a transfer veto) to other water-rights holders. The environmental review process, including citizens' rights to sue, also extends weak rights to capital to local community members and other interested parties. Using procedures for environmental impact reporting and the courts, third parties can exert some control over water-transfer decisions, which is part of the right to capital. Second, rights could have been withheld from individuals or public agencies by the state. The right to destroy water (by polluting it to the extent that no other use is possible), for example, is not generally granted to water-rights holders but requires special permission.

Potential for a Water Market to Adapt in the Wrong Direction

A third potentially negative characteristic of market-based water allocation involves a quality of markets typically considered a virtue: the market's ability to adapt to change. In 1945, Austrian economist F. A. Hayek published a seminal discussion of how price signals enable distant parties with detailed knowledge of their local circumstances to coordinate efficiently a region's economic activity. Markets utilize the knowledge of all their participants and adapt to change without any centralized planner being in charge.

This characteristic of markets is not always in a society's interest. In the case of water, a market could move a society's relationship with water in a direction the society would not have preferred had it been given a choice. For example, numerous independent and relatively small sales of water could over time severely deplete the resources of some regions of a state, severely affecting local economies and limiting opportunities for future growth.

California's Yolo County, especially its western side, is just such a candidate for missed opportunities for economic growth. With the exception of Cache Creek, western Yolo County has little surface water or groundwater available, and there are no substantive conveyance facilities from the nearby Sacramento River. With little water available for economic activity, it is unlikely that towns and industry there will amass the financial resources needed to compete with the San Francisco Bay Area and other neighbors for market-allocated water. Thus, market-based water allocation could cause long-term relative underdevelopment of western Yolo County.

Similarly, buyers could unexpectedly be concentrated in certain regions of the state, with unintended environmental and land-use consequences. For example, a market for water could move suburban sprawl onto prime farmland and into valued open spaces. With a decentralized reallocation mechanism, there is no opportunity for collective choice and no moral accountability for the results. Since water allocation undoubtedly has collective importance in California (and elsewhere), the adaptive qualities of a market could lead the state in otherwise undesirable directions.

So How Good an Idea Is Market-Based Reallocation of Water?

The possibility of a water market adapting to supply and demand signals in ways contrary to public goals for water reallocation has, to some extent, been reduced by the general policy of reallocating water from agricultural use to urban and environmental uses. The political nature of the reallocation process described earlier exerts a countervailing force on the potential negative effects of markets. Even if the specifics are not known, one can generally guess which interests will gain water and which will lose water.

This concern about adaptation is further reduced by the protection of some publicly valued uses of water through such means as minimum flow requirements and laws restricting the conversion of prime farmland. Further, to the extent that proposed transfers are evaluated for their consistency with the public's larger vision of water reallocation, some control can be exerted over specific regional effects.

One consequence of the evolution of property rights to water over the twentieth century is a reduction in security of ownership, which increases the cost of transacting and makes buyers less willing to purchase the resource. As a result of this history, market advocates today face an institutional-design choice. On the one hand, they can attempt to reconcentrate property rights enough to make market-based transfers less costly and more desirable to buyers. Proponents of this approach would argue that the benefits of a market-based reallocation would outweigh the costs to those who lose their current claims. This is the general approach taken in California's Model Water Transfer Act, described in chapter 2.

On the other hand, advocates can attempt to fashion institutions that limit the scope of market-based trading enough that the protections provided to nontraders by diffuse property rights can be largely retained in the new reallocation system. Proponents of this approach would be required to show that a limited market is sufficient to meet the state's reallocation goals. Appendixes 1 and 2, taken together, point out that the former approach—reconcentrating water rights—has been ineffective in creating a water market. The latter approach, however, remains largely unexplored. Subsequent chapters examine the task of generating a proposal for water reallocation that is consistent with the historical evolution of water rights in California and that minimizes the potential adverse consequences of a market-based mechanism.

Notes

1. Rural regions of California and certain classes of farmers, such as family farmers, may not be able to compete for water on the basis of price, and this may exclude them as an end point of water reallocation. There is little support for any kind of market mechanism from these interests.
2. *McDonald & Blackburn v. Bear River & Auburn Water & Mining Co.,* 13 Cal. 220, 232–233 (1859).
3. *Thayer v. California Development Co.,* 164 Cal. 117, 128 P. 21 (1912).
4. 1913 Cal. Stat. 586, sec. 12, at 1018.
5. *Herminghaus v. Southern Cal. Edison Co.,* 200 Cal. 81, 252 P. 607 (1926).
6. *Joslin v. Marin Municipal Water District,* 67 Cal. 2d 132, 60 Cal. Rptr. 377 (1967).
7. *United States v. State Water Resources Control Board,* 182 Cal. App. 3d 82, 227 Cal. Rptr. 161 (1986).
8. *National Audubon Society v. Superior Court of Alpine County,* 33 Cal. 3d 419, 658 P.2d 709, 189 Cal. Rptr. 346 (1983).
9. 1931 Cal. Stat. 1514, codified as Cal. Water Code sec. 10505 (West 1971 & Supp. 1985).
10. 1933 Cal. Stat. 2643, codified as Cal. Water Code sec. 11460 (West 1971 & Supp. 1985).
11. Cal. Water Code secs. 12930–12944 (West 1971 & Supp. 1985).
12. *Acton v. Blundell,* 52 Eng. Rep. 1223 (1843).
13. *Katz v. Walkinshaw,* 141 Cal. 116, 74 P. 766 (1903).
14. *Orange County Water District v. Farnsworth,* 138 Cal. App. 2d 518, 292 P.2d 927 (1956).
15. *Alameda County Water District v. Niles Sand & Gravel Co., Inc.,* 37 Cal. App. 3d 924, 112 Cal. Rptr. 846 (1974).
16. *City of Los Angeles v. City of San Fernando,* 14 Cal. 3d 199, 123 Cal. Rptr. 1 (1975); *Nebraska v. Wyoming,* 325 U.S. 589 (1945).

CHAPTER 4

Lessons from California's Experience with Short-Term Water Markets

"[And so] a lot of people weep
To think that water can't be cheap."

—from "The Cost of Water," a poem posted on the
wall of a suburban California water agency

California's lack of a market for long-term interregional rural-to-urban transfers of water stands in sharp contrast to the presence of numerous smaller-scale markets that handle short-term transfers. Short-term transfers involve the sale of water rights for a period of one year or less; long-term markets involve any transfer longer than one year. Short-term markets therefore do not present a threat of long-term reallocation of water away from farmers to urban regions and environmental uses. Their purpose is related to seasonal economic needs and opportunities of farmers and other market participants.

Seeking Lessons for Institutional Design of Markets

Short-term water markets clearly are functional. Farmers have been active sellers and buyers of water for decades within their individual water districts, with total transactions numbering in the thousands. Just as farmers trade water within districts, rural water districts also frequently trade water among themselves. For example, districts served by the Central Valley Project engaged in 1,200 short-term transfers from 1981 through 1988, ranging in size from a few to more than 100,000 acre-feet (af). The Westlands Water District in the San Joaquin Valley has a staff that coordinates numerous water transfers and exchanges on behalf of its farmers. In the case of interregional transfers using State Water Project facilities, roughly 6 million acre-feet (maf) changed hands from 1982 through 1993, nearly all on a short-term basis. This volume of trading provided a significant amount of flexibility in water allocation (Rosegrant 1995).

Among urban districts, many have participated as buyers in the state-operated Drought Water Banks, including, in 1991, the Contra Costa Water District, the Alameda County Water District, the Santa Clara Valley Water District, the city of San Francisco, and the Metropolitan Water District of Southern California (MWD). Groundwater leases also have been actively traded on a short-term basis in the Los Angeles Basin area since the 1960s.

Short-term water markets are effective allocation mechanisms. Appendix 3 describes two tests to which any proposed policy for reallocating water should be put. The first is a feasibility test: to what extent does it achieve the policy goal that motivates its consideration? The second is an appropriateness test: to what extent is the proposed policy consistent with society's values, narrowly or broadly construed? To be adopted, a proposal must show itself to be superior to other proposals and, at the least, consistent with minimum social standards. With their extensive trading activity and long history, short-term agricultural-sector markets for water clearly have passed both the feasibility test and the appropriateness test. They permit multiple transactions on preagreed terms (i.e., each new transaction does not have to be completely renegotiated), and prices emerge that are linked to prevailing supply and demand. They are long-standing and non-controversial. The Drought Water Bank also has been effective, but it has not been without controversy.

What accounts for these performances, given that long-term trading remains stalled? This chapter examines existing short-term transfer institutions in California for insights applicable to the design of long-term market institutions. Details are provided about two existing short-term markets, the state-sponsored Drought Water Bank and the water-trading activities of the Westlands Water District, located on the western side of the San Joaquin Valley. Then short- and long-term water markets are analyzed for their similarities (of which there are few) and differences (of which there are many).

The Drought Water Bank: Relief for South-of-the-Delta Water Users

In 1991, California found itself in a fourth consecutive drought year; the drought was the longest since the one in 1928–1934. Storage in the state's major reservoirs had dropped to 54 percent of average; five counties had declared drought emergencies; and for one county, Santa Barbara, the governor had declared a state of emergency. The State Water Project, California's second largest provider of water, announced in February that it would deliver only 10 percent of requested water to urban areas and none to rural areas. The Central Valley Project, the state's largest provider, planned to deliver only 25 percent of requested water to its urban and agricultural users, though subsequent deliveries would be higher for urban areas. The drought resulted in severe rationing in urban areas, expansion of groundwater pumping by Central Valley farmers that would ultimately cause an overdraft of 11 maf, and heavy stresses on fish and wildlife. According to one estimate, by 1991 one-third of all trees in the Sierra Nevada were either dead or dying from lack of water.

As dry winter months passed, urban and rural water agencies in the southern and coastal regions of the state were stirred to action. Many were scouring water agencies in the eastern and northern mountain ranges in an attempt to secure additional supplies. For example, the San Francisco Public Utilities Commission purchased roughly 13,000 af from Placer County and the Modesto Irrigation District in October and December 1990, and then in January 1991 it granted its general manager new authority to spend $4 million on additional short-term water purchases. The commissioners anticipated a purchase of 20,000 af at a cost of $200/af. Other major water districts searching for sellers in late 1990 and early 1991 included MWD, the Westlands Water District, and districts representing Contra Costa County, the Tulare Lake region, and Kern County. Soon, these and other prospective buyers all found themselves knocking at the same few water-rich doors in northern and eastern California. They quickly realized that it would be in all their interests to form a unified buying strategy. This strategy manifested itself in the Drought Water Bank.

Authorization to develop a water bank grew out of recommendations made by a governor-appointed Drought Action Team, and the California Department of Water Resources (DWR) was given operational responsibility. One of the DWR's first actions was to form a Water Purchase Committee that included representatives of governmental agencies and potential buying agencies. This committee developed a standardized contract for water sales to the bank as well as a standard offer to sellers of $125/af. This price was meant to yield to farmers a net income similar to what they would have earned on the farm plus an incentive to participate in the program. It was not meant to reflect the scarcity value of water during an extended drought. With the addition of costs of administration, monitoring, conveyance, and carriage water (outflow losses), the price to buyers was set at $175/af at the State Water Project's Delta Pumping Plant. Buyers would incur additional costs in moving the water from the plant to their own systems.[1]

At $125/af, sellers did come forward, and in less than two months, more than 300 contracts were in various stages of negotiation. By early June, most of the contracts had been processed. Because it was a drought period, there was plenty of capacity available in conveyance facilities. This eased the DWR's and the buying districts' conveyance challenges. In all, 821,000 af of water was sold to the bank. Farmers made water available to the bank by fallowing land (51 percent of the total), by replacing surface supplies with groundwater supplies (32 percent), and by transferring water stored in reservoirs (17 percent). Forty percent of all deliveries resulted from fallowing in the Sacramento–San Joaquin Delta region; the crops forgone, in terms of acreage, were most often corn and wheat. Total fallowed acreage was 166,000 acres, or slightly more than 2 percent of all farmland in the state.

But something unexpected happened. The anticipated demand simply did not materialize (Howitt, Moore, and Smith 1992). The DWR had conducted surveys of the critical needs of urban and rural water districts. Critical needs would be validated by the DWR only if buyers built into their calculations full utilization of existing supplies and put in place a stringent water conservation program. Table 4.1 lists purchasers' estimates of their critical needs as of April 1, 1991. In all, the eighteen purchasers indicated an unmet need for nearly 500,000 af through the approaching summer months. Given the nature of the survey, these figures were construed as minimum amounts to be purchased, so DWR officials purchased 820,000 af. However, less than 400,000 af was eventually resold (see table 4.2), with an additional 165,000 af dedicated to meeting Delta water-quality requirements. The unsold supplies remained in reservoirs, designated as "carryover storage." The amount actually purchased was one-fifth less than the surveys had indicated would be needed, and one-third fewer buyers ultimately participated in the market. Among the possible reasons for the reduction in demand were an unusually wet March, effective conservation programs carried out in urban and agricultural areas, and, possibly, the cost of the water. A rural official said that one reason why water went unsold was that

TABLE 4.1

Estimates of 1991 Critical Water Needs in California as of April 1, 1991

Supplier	Amount (acre-feet)	Percent of Total
Alameda County Water District	15,000	3.0
Alameda County Flood Control and Water Conservation District Zone 7	6,000	1.2
Castaic Lake Water Agency	2,830	0.6
Devil's Den Water District	2,000	0.4
Dudley Ridge Water District	10,000	2.0
Empire–West Side Irrigation District	2,500	0.5
Hospital Water District and others	10,000	2.0
Kern County Water Agency	100,000	20.1
Littlerock Creek Irrigation District	500	0.1
Metropolitan Water District of Southern California	215,000	43.2
Napa County Flood Control and Water Conservation District	700	0.1
Oak Flat Water District	2,000	0.4
Palmdale Water District	1,000	0.2
San Bernardino Valley Municipal Water District	3,000	0.6
San Francisco Public Utilities Commission	50,000	10.0
Santa Clara Valley Water District	39,500	7.9
Tulare Lake Basin Water Storage District	35,000	7.0
Westlands Water District	2,702	0.5
TOTAL	497,732	

Source: Howitt, Moore, and Smith 1992, exhibit B.
Note: Percentages do not add up to 100 due to rounding.

TABLE 4.2

Water Bank Allocations as of October 10, 1991

Agency	Amount (acre-feet)	Cost
Alameda County Water District	14,800	$2,590,000
Alameda County Flood Control and Water Conservation District Zone 7	500	87,500
American Canyon County Water District	370	64,750
Contra Costa Water District	6,717	1,175,475
Crestline–Lake Arrowhead Water Agency	236	41,300
Dudley Ridge Water District	13,805	2,415,875
Kern County Water Agency	53,979	9,446,325
Metropolitan Water District of Southern California	215,000	37,625,000
Oak Flat Water District	975	170,625
San Francisco Public Utilities Commission	50,000	8,750,000
Santa Clara Valley Water District	19,750	3,456,250
Westlands Water District	13,820	2,418,500
TOTAL	389,952	$68,241,600

Source: Howitt, Moore, and Smith 1992, exhibit B.

at $175/af, it was just too expensive for rural districts. That is, even in an extended drought, price matters.

Whatever the reasons for the unexpected drop in demand, taxpayers eventually lost $45 million as a result of water being purchased by the bank with public money but not resold. The media criticized the DWR for investing $45 million in unsold water rights during the drought, but the DWR responded sharply that carryover storage during a drought, however acquired, was a good thing.

There were other potential negative effects, and possible mitigations, for each source of supply: surface-water purchase, groundwater substitution, fallowing, and Delta pumping (see table 4.3). For example, participation in a groundwater-substitution purchase could result in overdraft, land subsidence, adverse effects on other pumpers (such as higher pumping-lift costs), degradation of water quality (e.g., higher concentrations of total dissolved solids), and reduction in nearby surface-water flows. The water bank's subsequent program EIR suggested mitigation techniques for each potential adverse effect. For groundwater substitution, prior evaluation of subsidence risks and water-quality effects was recommended, along with real-time monitoring of water quality and levels.

As 1991's summer growing season concluded and initial assessments

TABLE 4.3

Activities and Potential Adverse Effects Associated with California's Drought Water Bank

Activity	Potential Impact
Surface-water purchase	Decreased carryover
	Downstream temperature increase
	Loss of tailwater for wetlands
Groundwater substitution	Overdraft
	Subsidence
	Effects on other pumpers
	Degradation of water quality
	Effects on surface-water flows
Fallowing	Increased soil salinity due to high water tables
	Loss of food supply to wildlife
	Loss of sensitive plants in pastures
Delta pumping	Entrainment (trapping) of fish

Source: California Department of Water Resources, *State Drought Water Bank Program Environmental Impact Report* (Sacramento: California Department of Water Resources, 1993), table 1, p. xix, and discussion.

were made, the Drought Water Bank received both glowing accolades and sharp criticism. Douglas Wheeler, California's secretary for resources, hailed the bank as a "success story" that "laid important groundwork for California's water future" (Wheeler 1992, p. A-11). Economist Richard Howitt and his colleagues reviewed the bank and labeled it a "resounding success," a "significant achievement," and "remarkable," and they qualified any of their criticism as "trivial" compared with what would have happened without the bank (Howitt, Moore, and Smith 1992). Water buyers in the southern part of the state, many of whom were State Water Project contractors, seemed quite satisfied with the program. Timothy Quinn, leader of water-purchasing efforts at MWD, noted that the bank helped southern California avert "economic disaster" by allowing MWD to purchase 215,000 af of water. Indeed, Steve Macaulay, the energetic chief of the DWR's Water Transfer Office who spearheaded the formation and establishment of the Drought Water Bank, was subsequently hired away from public service to become the new head of the State Water Contractors, a nonprofit mutual benefit corporation that represents the interests of the California State Water Project Contractors.

Severe criticism was leveled at the water bank by representatives of northern California water agencies who believed that the arrangement had blocked the opportunity for a true market price to be realized. One unsigned parody of the bank, titled "Comrade Wilson's Water Plan" (after California's governor, Pete Wilson), likened the bank to an "outdated Soviet system for managing the economy" and stated that "the Water Bank actually killed an existing and flourishing market" and "stymied the efforts of willing buyers and sellers to reach mutually beneficial agreements." Criticism also came from regions that experienced economic loss associated with land fallowing. Yolo County's board of supervisors, noting that 13 percent of the county's farmland had been fallowed, estimated that 450 agricultural jobs had been lost and requested compensation of $129,000 from the DWR for the added county burdens on the General Assistance Program, Aid to Families with Dependent Children, and other support services. In a study of adverse effects on regions that sold water to the bank, the Rand Corporation estimated that farmers participating in the bank saw reduced operating costs of $17.7 million and reduced crop sales of $19 million, 20 percent less than what they would have sold had there been no water sales (Dixon, Moore, and Schecter 1993). According to the Rand team, these losses, though concentrated primarily in the regions where fallowing occurred, were offset by an infusion of wealth from the sale of the water. So although the program altered economic patterns compared with the normal variability in the farm economy, the study concluded that there was no detectable overall economic effect on the selling regions. However, in some water-losing communities, the bank did generate divisiveness between owners of local businesses and participating farmers, between landlords and tenants of fallowed farms, and between county governments and the DWR.

With the passage of time, some new conclusions can be posited with respect to Calfornia's first Drought Water Bank. It undoubtedly increased the feasibility of short-term water transfers, and it paved the way for an unprecedented number of north–south transfers in a short period of time. It swept aside technical (conveyance) barriers to north–south transfers and accelerated the transfer-approval process. The DWR increased its visibility and its credibility and mobilized some of its brightest talent to create new procedures and to process transfers with lightning speed. More than 820,000 af of water was purchased in a matter of months, and more than 250,000 af passed through the ecologically sensitive, multijurisdictional Bay-Delta region. The close cooperation and timely action of the DWR, the Bureau of Reclamation, and buyers and sellers of water may have helped lay the groundwork for the subsequent state–federal cooperation in the immense San Francisco Bay/Sacramento–San Joaquin Delta restoration project known as the CALFED Bay-Delta Program.

An environmental benefit of the Drought Water Bank was the requirement that all transfers passing through the Delta include carriage water, or planned outflows to balance the additional pumping in the southern part of the Delta. The drought was especially stressful for the Delta ecosystem, and the carriage water provided instream benefits. In 1991, 165,000 af of water, or roughly one-fifth of all water purchased, served as carriage water.

On the other hand, far from being a proof of the water-market concept, the 1991 Drought Water Bank highlighted limitations with which future markets would have to contend. Regions where fallowing had occurred, such as eastern Yolo County, bore the brunt of adverse economic effects, and residents rose in protest. In the northern and eastern parts of the state, a practice that was equally disturbing, if not more so, was the use of groundwater substitution where aquifers were already in a state of overdraft. Residents saw groundwater substitution as an assault on their own drought-protection resources as well as a mechanism by which one farmer could profit (through the sale of surface rights) from exploitation of a shared resource (groundwater) for irrigation. Further, groundwater supplies pumped from wells located near rivers might already be hydrologically connected through underground flows to the Delta system. If so, the groundwater-substitution method would not actually introduce new water into the system.

The following March, with the drought continuing, the 1992 Drought Water Bank opened. Responding to the foregoing concerns, the DWR's announcement of the bank's opening addressed source-of-supply concerns (DWR 1992, p. 1):

> Whenever possible, [the DWR], which operates the program, will buy water without requiring that land be taken out of production. . . . The Bank also will strive to avoid taking water from overdrafted ground water basins.

By the end of September, when purchase requests finally abated, less than 1,500 acres of land had been fallowed to provide water to the bank (compared with 420,000 acres the previous year), and groundwater-substitution sales were being carefully monitored with meters on pump-discharge lines. The purchase price from sellers of water had been lowered to $50/af and the price to buyers of water was $72/af. A total of 193,000 af of water was purchased and 159,000 af delivered. The difference, 34,000 af, was carriage water. This time, the DWR had not purchased water in advance of sales, so it was not stuck with unsold quantities at the end of the summer. The Drought Water Bank reopened in 1994, and it was prepared to open in 1995, but wet weather ended preparations.

To conclude, California's Drought Water Bank experience can be evaluated from two perspectives. In a positive light, it demonstrated that a centrally organized system of short-term transfers can play a role in mitigating the adverse effects of drought. In a less positive light, it demonstrated the political and economic power of water districts along the southern coast, in the San Francisco Bay Area, and in the San Joaquin Valley, which were able to employ the DWR as their low-price, monopsonistic buyer of water from northern and eastern California. The Drought Water Bank experience further demonstrated (if further demonstration was needed) that water use and water transfers raise sensitive local concerns throughout the state. Even when legal rights to water are defined to provide an opportunity for marketlike transfers, local pressure can attenuate or stop proposed transfers. In the case of the Drought Water Bank, land fallowing and groundwater substitution bore the brunt of the criticism.

Westlands Water District: Local Efficiency through Short-Term Markets

Located on the western side of California's San Joaquin Valley, the West-lands Water District is the largest agricultural water-delivery agency in the United States. Westlands covers an area fifteen miles wide and seventy miles long, including 600,000 acres in both Fresno County and Kings County. Roughly 530,000 acres are irrigated. The district produces half the coun-try's supply of garlic, one-quarter of its processing tomatoes, and, for two months in spring and fall, 95 percent of the country's supply of lettuce. One-quarter of all cotton produced in California comes from Westlands; California is one of the country's leading cotton producers. In recent years, the active involvement of Westlands in a number of state water-policy issues has increased its public visibility.

In terms of water delivery and use, the district's contract with the fed-eral Central Valley Project (CVP) calls for delivery of 1.15 maf/year to 650 farmers. Prior to 1989, in some years Westlands had been able to purchase additional water, called "interim water," from the CVP, boosting its average annual CVP deliveries to 1.23 maf during 1979–1989. Westlands also can draw as much as 200,000 af/year from an underlying aquifer on a sustain-able basis. According to the district, its normal-year total supply falls roughly 20 percent (350,000 af) short of its normal-year potential cropping demand.

Since the mid-1980s, both natural and regulatory events have tightened the water-supply situation in Westlands. The same six-year (1987–1993) drought that led to the establishment of the Drought Water Bank also reduced CVP supplies to Westlands. Other blows to the district's water sup-plies included the signing by President George Bush of the 1992 Central Valley Project Improvement Act, which reallocated 800,000 af of CVP water to environmental purposes, and the listing of winter-run chinook salmon under the Endangered Species Act (ESA). The listing increased pressures for greater instream flows in the Sacramento River, potentially at the expense of storage for CVP contractors. ESA listing of the delta smelt added new pressure to reduce water diversions from the Delta south into aqueducts owned by the state and federal governments.

Westlands has responded to the actual and potential long-term tight-ening of its water supply by adopting new conservation techniques and by attempting to diversify its sources of water. Conservation technologies adopted include drip irrigation and microirrigation practices. In terms of diversification, the district has purchased almost 1.3 million additional af of water since 1989, all of it on a short-term basis. Wetter weather since 1994 has restored its annual CVP allocation to at least 90 percent of the contracted level, with total supplies to Westlands from all sources in 1996–1997 reaching 1.4 maf, the highest level since before the drought.

Another method of dealing with water scarcity has been intradistrict transfers. According to water marketing analyst Janis Carey (Olmstead 1997), during the 1994–1995 crop year, roughly 3,500 water transfers were made, involving 380,000 af of water. That means that more than one-third of all water delivered ended up somewhere other than where the district originally expected to send it, and 3,500 different reallocation decisions were made in the course of just one year. Some of the transfers were carried out between farmers; in others, individual farmers moved planned deliveries from one parcel to another on the same farm. The magnitudes of both internal transactions and quantities transferred indicate that Westlands farmers were carefully monitoring the prevailing commodity prices, the types and prevalence of pests, the health of their planted fields, and other factors and making ongoing calculations with respect to economic use of water. In other words, in terms of economic output, the farmers were making the most of the water they had through real-time reallocation decisions throughout the year.

The typical method for transferring a water allocation in the Westlands Water District involves farmers contacting each other, by telephone or otherwise, and coming to an agreement. Paperwork is then prepared and submitted to the district's Customer Accounting Department, which records the transfer and alters billing records. Compensation for the transfer is handled by farmers on a bilateral basis. In 1996, a computerized bulletin-board service called WaterLink was established, allowing farmers to post requests to buy and sell water at specific prices. It also provides an e-mail link so that farmers can contact one another directly to negotiate agreements. In its first year, only a handful of transfers could be credited directly to the new computer system. Possible explanations for the low rate of utilization include reduced interfarm transfers during a year of high CVP allocations and a fear among farmers that posting information on willingness to buy and sell water at a specified price would put them at a competitive disadvantage. It also is possible that with 3,500 transfers already taking place in one year, there was no communication problem that needed fixing. That is, farmers had already developed effective noncomputerized means for sharing information on prices and quantities and reaching agreements on transfers.

Environmental effects directly attributable to short-term transfers in

Westlands appear to be minimal and perhaps positive. Short-term transfers give farmers the flexibility to reduce groundwater pumping, rotate crops, and fallow fields in ways that help preserve the long-term health of the soil (as opposed to having fixed water allocations available to each field). Short-term transfers also give farmers flexibility to deal with perched water tables (when water no longer percolates down due to full saturation of the ground) and saline intrusion into root zones.

To summarize, the farmers of the Westlands Water District have developed an effective means of moving water within the district and achieving optimal economic usage. The transfers take place under conditions of potential long-term reductions in water deliveries as well as annual shifts in the amount delivered by the CVP. Each year, hundreds of thousands of acre-feet of water are moved and thousands of transfers take place, a practice that undoubtedly contributes to the financial success of the district's farmers. The physical capacity to move water is not a constraint within the district. Environmental effects of the short-term market appear to be relatively minor but positive.

Similarities and Differences in Short- and Long-Term Markets

As stated at the beginning of this chapter, the policy purposes of short-term water transfers differ from those of long-term transfers. Although short-term transfers increase the marginal efficiency of water use, they do not play a role in long-term reallocation of California's waters. They therefore avoid much of the controversy that surrounds proposals for long-term rural-to-urban transfers. Appendix 4 gives two examples of controversial long-term proposals and a possible economic explanation for individual farmers' opposition to such transfers. Short- and long-term transfers have many other similarities and differences as well.

In terms of similarities, the economic reasoning behind both short- and long-term markets is the same: willing sellers and buyers of water can be made better off if the water is available for transfer. Other similarities include a requirement that water transferred be water actually in use, or "wet" water, as opposed to unused water rights, or "paper" water. Short- and long-term water transfers also involve moving or extending the place of use and perhaps changing the purpose of use. Only the duration of the change differs.

If the trading *experiences* (i.e., the steps involved in transacting) with respect to short- and long-term agreements were identical, one would be hard-pressed to find reasons why one type of market thrives while the other founders. But the experiences clearly are not identical, especially in the negotiation, closure, and execution of agreements. From the agricultural seller's perspective, a short-term sale can be easily linked with available supplies and conveyance, seasonal commodity prices, weather conditions, pest prevalence, and other temporary factors, whereas long-term sales are linked with long-term choices related to cropping patterns, capital investment in irrigation technology, fallowing, or even business closure. (Some of the long-term considerations just mentioned might also lead a farmer to consider a short-term sale or lease of water rights.)

From a buyer's perspective, a short-term transfer can be financed by means of current assets or a short-term loan. But for a more expensive long-term purchase, buyers must arrange financing that is equitable to both current and future generations of water users. That is, if the total cost of the transfer is paid up front, today's residents are subsidizing tomorrow's users. Resistance to such financing plans is likely to grow when future users are projected largely to be newcomers to a district and not primarily the progeny of existing residents. On the other hand, if everything is financed in anticipation of future payments by future users, the financial stability of the local water agency hinges on growth projections that may prove to be inaccurate.

Another difference between short- and long-term transfers is that short-term transfers typically meet a buyer's well-defined, near-term *water-quantity* need, whereas long-term transfers often must also meet a *water-reliability* need. That is, cities commonly seek water rights to ensure that they have future drought supplies as well as water available to cover a forty-year high-growth scenario. In their competition to attract investment in desirable industries (such as high-tech research and development or manufacturing), cities want sufficient water rights to assure potential investors of guaranteed water supplies for decades into the future. The distinction between quantity and reliability is linked to availability of conveyance facilities. In California, conveyance of north–south transfers through the Delta can typically be arranged for a single season or year, but a long-term commitment to new conveyance patterns through the Delta is extremely difficult to procure.

Since the purposes behind efforts to acquire short- and long-term rights differ, the organizational approach to acquiring the rights also differs. More data from different sources must be compared when quantifying and costing out long-term reliability needs than when quantifying and costing out seasonal adjustments.

Environmentally, there are significant differences between short- and long-term transfers. In California, both ecosystems and social systems are attuned to temporary fluctuations, such as those caused by the state's periodic droughts. A short-term loss of water from a region due to a one-year transfer can be compared to a drought. In subsequent years, the region bounces back. A combination of prudent management of a short-term transfer program and intermittent transfers from numerous sources is likely to minimize the local social and environmental effects of short-term transfers. By contrast, a long-term loss of water has the potential to alter a landscape. In regions where farming has been carried out for a century or more, both ecological and social systems are attuned to historical cropping patterns. Some species depend on grazing in grain fields; others depend on runoff of tailwater (water that accumulates at or spills over the end of a field following flood or furrow irrigation) to feed wetlands. Populations of native species may rebound if irrigation-based farming practices cease. Altered timing, quality, or quantity of instream flows along ecologically important waterways could also have environmental effects.

Thus, many contrasts exist between short- and long-term markets, not least of which is the distinction between water as an input to the agricultural economy (short-term markets) and water as a statewide commodity to be reallocated from farms to cities (long-term markets). These contrasts, however, do not nullify the value of examining short-term markets for features that could inform the design of a long-term reallocation strategy. Short-term markets (especially agricultural-sector markets) have proven to be feasible and appropriate, and parts of their institutional design might be transferable to a long-term setting. These insights, combined with lessons from California's three leading examples of long-term transfers (chronicled in chapters 5–7), can provide guidance in designing effective institutions to reallocate California's water over the long term.

Note

1. The $175 cost was less than the $200/af that the San Francisco Public Utilities Commission had anticipated spending one month prior to formation of the bank.

CHAPTER 5

Shotgun Wedding: The Imperial Irrigation District–Metropolitan Water District Water Conservation Agreement

"[T]he Manure Mafia . . . [is] trying to sell off the water. . . . They call it conservation but this is a big lie."

—letter to *Imperial Valley Press* from Ben Yellen, M.D., March 21, 1989

Of all the long-term rural-to-urban water transfers that have taken place in California, the transaction between the Imperial Irrigation District (IID) and the Metropolitan Water District of Southern California (MWD), two titans among water institutions, may be the most widely cited and least understood. It is the great water-market success story to which everyone points, but the deal was exceedingly difficult to make happen. For that reason and others, it is not a model for western-state water reallocation. However, lessons do emerge from it that can inform efforts to create feasible and appropriate long-term market institutions. These lessons are presented in the concluding section of this chapter.

Regional and Institutional Backdrop

Both the IID–MWD transfer and the temporary agreement between MWD and the Palo Verde Irrigation District discussed in the next chapter

involve the westward transfer of Colorado River water. Since both transfers take place in generally the same hydrologic, institutional, and legal settings, the following remarks serve to introduce both IID's and Palo Verde's
deals with MWD.

Hydrology

The Colorado River, with headwaters in the mountains of Wyoming,
Utah, and Colorado, serves as the boundary between California and Arizona on its southwestern passage to the Sea of Cortés (Gulf of California).
In the twentieth century, geologist and explorer John Wesley Powell trav-
eled the wild Colorado by wooden raft, a trip that lasted several days. We
will now take a quicker tour.

As shown in figure 5.1, south-flowing floodwaters from the lower Colorado River are captured and stored or diverted by a series of dams, including Parker Dam and Imperial Dam. Reservoirs along the river can store 40
million acre-feet (maf) of water. The Colorado River also forms the Ari-
zona-Mexico border for roughly thirty-five miles before crossing into
Mexico.

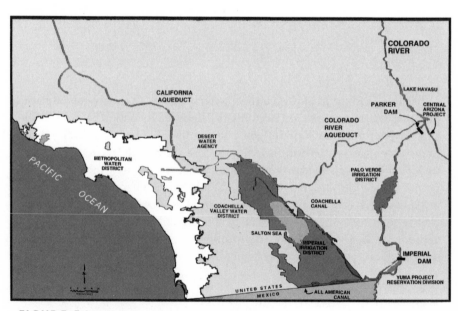

FIGURE 5.1
California's Connection to the Lower Colorado River
Source: Metropolitan Water District of Southern California.

Just below the tri-state (Nevada, Arizona, California) intersection is Lake Havasu, which lies behind Parker Dam. At the lower end of Lake Havasu, on the western side, is the intake for the Colorado River Aqueduct. This aqueduct, built, owned, and managed by MWD, carries water 242 miles west from the Colorado River to the urban regions of the southern coastal plain, including Greater Los Angeles and San Diego. The Colorado River Aqueduct has the capacity to move 1.3 maf/year and typically moves 1.2 maf/year from east to west. This is the portion of MWD's water with the high content of dissolved solids.

The hot, arid region in California's southeastern section known as the Sonoran Desert achieves temperatures considerably higher than 100 degrees Fahrenheit in summer and 70–80 degrees Fahrenheit in winter, with an average annual rainfall of only 2.2 inches in the city of Blythe. At night, the skies are a stunning splash of stars.

From Parker Dam, moving south along the river about sixty miles, one comes to the Palo Verde Valley on the western (California) side. The valley is nine miles wide and thirty miles long, ranging in elevation from about 290 feet at the northern end to about 220 feet at the southern end. Prior to the 1935 taming of the lower Colorado River's floodwaters with the completion of Boulder Dam (now Hoover Dam), the valley was regularly inundated, receiving not just the waters but also the nutrient-rich sediments carried by the Colorado. Total dissolved solids during spring floods range between 200 and 300 milligrams per liter; during periods of low flow, they have reached 1,500 milligrams per liter. In recent years, salinity measured in parts per million (ppm) has stayed within the 700–800 range at the Palo Verde Diversion Dam, rising to 800–900 ppm at Imperial Dam and then to 900–1,000 ppm at the Mexican border. The valley's soils are alluvial (good for crops), ranging in texture from fine-grained clays to silty loams to light, sandy soils, with the predominant soil being a sandy loam. Soil salinity is variable, with the highest concentrations in bottomlands and where gullies formerly returned floodwaters to the river. The entire valley is underlain with permeable sand at shallow depths. In short, it is a very fertile valley, and it has been productive farmland for more than a century.

Palo Verde groundwater is linked hydrologically with the Colorado River and generally flows from north to south. In recent years, average groundwater elevation has fluctuated in the range of 8.8 to 10.4 feet below the ground surface, with a slight deepening trend since 1989. As of 1989, there had been no archaeological excavations in the Palo Verde Valley region.

The Palo Verde Diversion Dam moves water at the rate of some

900,000 acre-feet (af) per year into the canals and laterals of the Palo Verde Irrigation District (PVID). According to a pamphlet regularly updated by PVID, the combination of hot, dry climate, year-round cropping, minimal slope, and salty soils means that a great deal of water is applied to the soils—as much as 7.7 af per cropped acre per year (PVID 1994). Because both the aquifers and the canal system are linked with the river at the lower end of the valley, return flows to the river are substantial: about 4.1 af per cropped acre per year, which nets an average consumptive use of about 3.6 af per cropped acre per year.

Let us stop our tour momentarily to note the beginnings of the attraction of transferring Colorado River water from farmers to cities. Farmers in the Palo Verde Valley apply an average of 3.6 af per cropped acre per year, or enough for roughly ten urban families. Thus, if a farmer fallows just one acre, enough water is liberated to serve the needs of ten families. Families count water use in gallons; farmers think in terms of acre-feet. Families in cities pay approximately 100 times more for water than Palo Verde farmers do. A market opportunity seems to exist here.

Moving farther south, just a few miles above the international boundary is Imperial Dam, which diverts water into the All-American Canal. The All-American Canal was named for the fact that although it hugs the U.S.–Mexico border, it never crosses it. It moves Colorado River water west (more than 3 maf/year—usually much more than is left in the river) for eighty-two miles to the Imperial Irrigation District, which lies south of the Salton Sea.

The Coachella Valley Water District (CVWD), located on the northern end of the Salton Sea, also receives its water from the All-American Canal, which feeds the north-running Coachella Canal. This hydrologic connection between the Coachella Canal and the All-American Canal, combined with the intricate water-rights arrangements discussed later in this chapter, became the biggest stumbling block for MWD and IID as they attempted to craft an agreement. In essence, MWD and IID had to accommodate CVWD's needs; otherwise, CVWD could have blocked the deal.

The geography of the region around the Salton Sea makes it a natural sink for Colorado River floodwaters. That means that when the Colorado's course is diverted to the Salton Sea, as has happened occasionally through the ages, there is no place left for the river to flow. The Salton Sea is the region's geographic low point, so it simply keeps filling up until the next storm or other natural cataclysm reroutes the Colorado River back to the Sea of Cortés and the Pacific Ocean. The natural terrain of this region prohibits IID and CVWD from providing reusable return flows to the Colorado River. The Salton Sea was filled between 1905 and 1907, when the Salt and Gila Rivers, tributaries to the Colorado River, flooded and knocked out the headgates of an older canal that conveyed water to the Imperial Valley through Mexico.

Since the building of the All-American Canal, the Salton Sea's level has been determined primarily by a balance between irrigation runoff and evaporation. In the case of IID, two small rivers, the New and the Alamo, receive farm runoff from 1,460 miles of ditches draining roughly 440,000 acres of farmland, and then they flow north into the Salton Sea. In years when IID takes 2.6 maf from the All-American Canal, delivery losses reach an estimated 300,000 af, and on-farm losses reach 500,000 af (enough for as many as 2 million families). On-farm consumption amounts to the difference: 1.8 maf. The 500,000 af of annual IID farm losses amounts to roughly 7 percent of the Salton Sea's 7.3-maf volume. Because the only outflow from the sea is evaporation, the level of salinity has slowly increased through the years; in 1999 the salinity was 44,000 parts per million (ppm), which exceeds that of seawater, at 35,000 ppm.

The Salton Sea is home to many waterfowl, among them the brown pelican, listed as an endangered species. It also has been stocked for sport fishing, which supports the local economy.

Through the years, IID has faced two primary concerns regarding the Salton Sea: pollution runoff in agricultural drainage and the rising elevation of the sea. Selenium levels in both agricultural drainage water and benthic (lake-floor) sediments have exceeded state advisory levels. Also, the New River, which runs north from Mexico, carries untreated sewage, heavy-metal contamination, and pesticide residues released south of the border through IID, where the river's water is mixed with IID's drainage water and deposited into the Salton Sea. Although IID is not held responsible for pollution of the New River originating in Mexico, its additional effluents receive scrutiny in part because of the New's reputation for heavy pollution.

The steadily increasing level of the Salton Sea, which rose from 250 feet below sea level in 1925 to 227 feet below sea level in 1985, has resulted in damage to property along its shores (primarily farm acreage), complaints from property owners, and public scrutiny of the water-use practices of IID.

Returning to the region's canals, because the All-American and Coachella Canals follow a gradual downward slope from the Colorado River, they can deliver water to the two irrigation districts solely by means of gravity.[1] Although this saves considerably on the cost of moving water, the two districts' reliance on gravity flow makes them extremely sensitive to the volume and flow regimes of the river and canal. The river is managed from Lake Mead by the federal Bureau of Reclamation, with which IID and CVWD place orders for water. There is both an annual water order, made in October, and weekly water-requirement orders. Once released from Lake Mead, the water takes five days to arrive at IID's three main canals. Let us turn now to the ways in which society has organized itself around these physical features.

Institutions

MWD, PVID, IID, and CVWD, along with the Yuma Project of the U.S. Bureau of Reclamation, are the primary consumers of California's 4.4-maf annual portion of Colorado River flows.

METROPOLITAN WATER DISTRICT OF SOUTHERN CALIFORNIA

MWD is a wholesale water agency that was established in 1928. MWD's original purpose was to build and manage the Colorado River Aqueduct, which began water deliveries in 1941. It subsequently contracted with the State Water Project for deliveries of water originating in the Sierra Nevada and other northern California mountain ranges. These deliveries commenced in 1972. Today, its primary purposes are to develop, store, and distribute water for municipal and domestic use to its twenty-seven member agencies. Nearly all of its water comes from the Colorado River Aqueduct and the State Water Project.

MWD's service territory covers 5,139 square miles, extending 200 miles along California's southern coast from Oxnard to the Mexican border, a massive territory compared with that of other urban agencies. The population in its service territory is expected to grow from over 16 million people in 1999 to more than 21 million by 2020.

On average, MWD fulfills about 55 percent of the water needs of its service territory. The region's total demand grew from about 2.8 maf/year in the 1970s to almost 4 maf/year in 1996. Some member agencies, including those in Beverly Hills, Burbank, Las Virgenes, and San Diego County, rely on MWD for more than 90 percent of their water. By 2020, territorywide dry-year demand could increase by as much as 1.5 maf over 1996 usage.

MWD has a fifty-one-member board of directors, with each member agency represented by at least one board member. A general manager serves at the discretion of the board. In 1990, MWD had 1,700 employees.

Some relatively recent legal decisions have weakened this huge area's claims to various sources of imported water. The U.S. Supreme Court's 1963 decision in *Arizona v. California* reaffirmed California's apportionment of Colorado River water at 4.4 maf/year, not the 5.1 maf/year the state had claimed.[2] California has a right to half the surplus flows of the Colorado River, and the state had argued that Gila River flows in Arizona should be counted as part of the surplus. The Court's decision further required that any current or future claims from Indian tribes be met from state portions. Although Congress subsequently gave California's portion priority over Arizona's, the decision put at risk roughly 1 maf of surplus water that California had been taking from the river for years. When the Central Arizona Project aqueduct began diverting water in 1985, MWD's ability to divert water was put at greater risk. MWD now has firm rights to 550,000 af/year of Colorado River water and relies on unused water from prior appropriators, unclaimed Indian water rights, and unused Arizona and Nevada water to make up the difference. Conveyance losses along the aqueduct reduce the amount delivered to the southern coast by roughly 50,000 af/year (MWD 1990, 58). In 1990, for the first time in history, the Colorado River was unable to meet the requirements of California users. In 1991, it happened again.

Other recent actions that threaten the southern coast's long-term water security include the Los Angeles Department of Water and Power's loss of diversion rights to feeder streams to Mono Lake[3] and the decades-long San Francisco Bay/Sacramento–San Joaquin Delta negotiations, which will directly affect, either positively or negatively, the south coast's ability to receive north–south transfers of water contracted through the State Water Project.

PALO VERDE IRRIGATION DISTRICT

PVID began operations in 1925, when separate irrigation, drainage, and flood-control districts were merged. Today, the valley portion of PVID contains 103,500 acres; an additional 16,000 acres are on the Palo Verde mesa, just west of the valley. The Colorado River forms the eastern and southern boundaries of the district, which has 250 miles of canals, 40 miles of which are concrete lined. An additional 140 miles of drainage channels lead back to the river.

More than half the acreage in PVID produces alfalfa, followed by cotton, Sudan grass (for pasture and hayseed), wheat, and fruits and vegetables. Alfalfa essentially is solid green milk: the water-intensive food source for livestock. The vast majority of PVID alfalfa is shipped west to feedlots closer to the populous coastal regions.

The district has a staff of approximately eighty-five people, a number that has been falling in recent years as irrigation management has become less labor-intensive. A general manager serves at the pleasure of a seven-member board of trustees. The board is elected by landowners within the district on the basis of one vote for each 100 units of district-assessed valuation; that is, the larger the landholding, the larger the number of votes. PVID appears to be financially solid, with per-acre bonded indebtedness of district lands ranking among the country's lowest.

IMPERIAL IRRIGATION DISTRICT

Although the heyday of its political and economic power has passed, IID, organized in 1911, remains one of the country's largest and most influential irrigation districts. It was Imperial Valley's congressman (and former IID chief counsel) Phil Swing who drafted and championed the 1928 Boulder Canyon Project Act, which authorized construction of not only Hoover Dam but also the All-American Canal. In 1996, IID had roughly 1,200 employees and ranked among the country's largest irrigation districts.

IID's service territory extends over 1 million acres and currently delivers water to 495,000 acres of cropland and lands with agriculture-related uses and 25,000 acres of towns and residence-oriented areas (parks, fairgrounds, etc.). With a baseline water demand of 2.8 maf/year, IID farmers in 1994 planted 36 percent of their acreage with alfalfa, 15 percent with Sudan grass, and 11 percent with wheat. The balance of the acreage was planted with a supermarket variety of garden crops, field crops, and permanent crops.

Water supply is just one of two activities (and the lesser in economic terms) in which IID engages. In addition, the district manages the generation, purchase, and delivery of electric power to 45,000 customers. Its 1992 peak load of 556 megawatts and its total delivery of 2.35 billion kilowatt-hours make it the sixth largest electric utility in the state. Compared with its Water Department, IID's Power Department has more than twice the assets and generates roughly four times the income. If IID's service territory of Imperial and Riverside Counties remains the fastest growing in the state in terms of population, the power-to-water revenue ratio should grow even further. Despite this disparity, IID still stresses its name's second word in terms of its public image, internal organization, and composition of board members.

COACHELLA VALLEY WATER DISTRICT

Formed in 1918, CVWD provides water and flood-control services to both agricultural and urban customers at the northern end of the Salton Sea. In 1994–1995, CVWD provided irrigation water to 65,100 acres in a total service area of 638,000 acres. It provided roughly 283,000 af of Colorado River water to its agricultural customers and 80,000 af of well water to the 161,000 people who reside in its service territory. CVWD has 474 employees.

In contrast to the situation in PVID and IID, CVWD growers have only 3 percent of their acreage in alfalfa. Sixty percent of the district's farmland produces table grapes, grapefruit, dates, oranges, and other fruits, and 31 percent produces a wide variety of vegetables. One reservoir, Lake Cahuilla, provides 1,300 af of storage for irrigators, less than a two-day supply. Historically, irrigation water arrived via the Coachella Valley Canal and urban supplies came primarily from wells. More recently, as farmers have switched to drip irrigation, they have begun to apply groundwater, which is less likely to clog the drip emitters because of its lower concentration of dissolved solids. This transition is creating a groundwater overdraft problem in the lower Coachella Valley. In the early 1990s, the CVID ranked among the country's top ten water districts in terms of both gross crop value and revenue per irrigated acre. (IID also ranked among the top ten in terms of gross crop value.)

Water Rights

Buried in the details of California's water rights to the Colorado River are the hooks that CVWD and other districts use to influence the transfer decisions of their neighbors. Key clauses of the transfer contracts make no sense if one does not understand the underlying rights, sometimes called "the law of the river," which are presented in this section.

The 1931 Seven-Party Agreement governs the use of Colorado River water in California, except for certain previously existing water rights. California's annual guaranteed allotment from the Colorado River is 4.4 maf/year.

Four agricultural agencies share the first three entitlements to California's portion: PVID, IID, CVWD, and the Yuma Project of the U.S. Bureau of Reclamation.[4] The order of entitlement begins with PVID, which, on the basis of Thomas Blythe's 1877 broad request to appropriate water, can take as much water as is needed to irrigate 104,500 acres in the Palo Verde Valley. Although PVID's right is unquantified, between 1966 and 1991 it consumed an average of 422,000 af/year (4.7 af per irrigated acre), with maximum use of 515,000 af in 1981. This unquantified first priority puts PVID in a class of its own: no one's rights are more generous or better positioned. Next is the Yuma Project Reservation Division's entitlement to such water as is required for beneficial use on a maximum of 25,000 acres. The Yuma Project generally consumes 70,000 af/year (Hyduke 1991), but it received as much as 92,000 af in 1989.

The third priority is shared by IID, CVWD, and PVID (the latter to provide water to its mesa lands). IID consumes approximately 2.8 maf/year, and CVWD receives roughly 350,000 af/year. Because of the cost of lifting river water onto the Palo Verde mesa, the mesa's marginal soil, and the availability of well water not historically linked to Colorado River flows, river-irrigated acreage on the mesa has declined, and PVID takes relatively little water in this priority. Although the Seven-Party Agreement specifies that this third priority is equally shared among the three agencies, the subsequent 1934 Agreement of Compromise between IID and CVWD gives IID first priority to water flowing in the All-American Canal for use "exclusively in [IID's] service territory."

The third priority is capped in this way: its holders can take as much as 3.85 maf/year minus what PVID and the Yuma Project consume from their first and second priorities. In effect, this sets a total consumption cap of 3.85 maf on the first three priorities.

The fourth and fifth priorities are both held by MWD and total 1.212 maf.[5] A sixth priority also is shared by IID, CVWD, and PVID for its mesa lands, totaling 0.3 maf/year. A seventh priority, for all remaining water available for use within California, is reserved for agricultural use in the Colorado River basin within California.

In years when Arizona and Nevada do not consume their full entitlements, California has the right to divert as much as one-half of that surplus. MWD has a surplus-water contract with the Bureau of Reclamation for a maximum of 0.18 maf/year.[6]

Readers with a mind for arithmetic may have noticed that the entitlement quantities in the seven priorities far exceed California's annual 4.4-maf share of the river. Under the appropriative doctrine, water returned to the river or not used by a senior appropriator may be used to fulfill a junior appropriator's share. In other words, it is all right to leave water in the river, and senior appropriators are expected to return surplus water to the river. MWD, as a junior appropriator, has a firm right to only 550,000 af/year because this amount falls under the state's cap of 4.4 maf/year. This amount could conceivably be reduced even further—perhaps by another 20,000 af/year—if all California Indian tribes with Colorado River rights took their full allotments. But if the agricultural agencies or other states do not consume their full entitlement, MWD has the right to take what the agencies leave—up to its 1.212 maf/year state limit as well as its 0.18 maf/year surplus limit—as constrained by the maximum flow volume (1.3 maf/year) of the Colorado River Aqueduct. Thus, only the first four priorities are certain of receiving Colorado River water in a normal year.

There is one other basic element of the water-rights situation. The rights described thus far are state rights based on the 1931 Seven-Party Agreement and written into the various entities' water contracts with the Bureau of Reclamation. Federal water rights also overlap the state rights. These are held by the parties that were drawing water from the Colorado River before construction of the major dams. With respect to federal rights, IID has "present perfected rights" to 2.6 maf/year. PVID and the Yuma Project also have federally recognized water rights, but MWD, which did not start diverting water until 1941, does not. The federal rights precede the distribution defined by the Seven-Party Agreement and essentially create for IID (and other similarly situated districts) a minimum consumption level to which it is entitled, even during a dry year when other appropriators are cut back.

That, then, is how the appropriative system works. Now let us look at a big caveat. The California Constitution (Article 10, Section 2) contains language stating that all water must be used in a "reasonable and beneficial" manner. That means that if it is shown that a rights holder—even PVID—is wasting water, hoarding it for speculative purposes, or otherwise engaging in unreasonable or nonbeneficial use, the rights can be taken away. (As discussed later, this was the shotgun at IID–MWD nuptials.) For farmers, nearly all traditional irrigation practices are considered to be "reasonable and beneficial," even though they may not conserve as much as they could, given recent innovations in irrigation technology. One controversial practice, however, involves the creation of excessive tailwater. Many parties believe that the creation of excessive tailwater is wasteful; if it is, farmers would lose their rights to such water, including the right to transfer it to other users.

The IID–MWD Negotiations

This section begins with historical background as to why IID and MWD were initially motivated to come to the table and then examines the first two distinct phases of negotiations: the ill-fated initial round, which collapsed in 1986, and the much-hailed second round, which culminated in an agreement at the end of 1988. The original accord, the resulting storm of criticism that sank the deal, and the subsequent third round of negotiations, which resulted in a new agreement, are then discussed. Following that is a review of round three, when two eleventh-hour agreements fundamentally altered the original accord but preserved the transfer as a whole and probably preserved the legal autonomy of IID.

What Originally Motivated the Parties to Negotiate

Water-conservation pressure on IID was building during the 1970s. A number of landholdings were flooded by rising waters from the Salton Sea, which is fed by agricultural runoff from IID farms. The landowners successfully sued the district for damages, and IID raised water tolls slightly to cover these costs. However, in 1984, John Elmore, a farmer with large holdings on the southwestern edge of the Salton Sea, filed a lawsuit asserting that IID was violating the state constitution's requirement that all water be used in a reasonable and beneficial manner.[7] He argued that the flood damage to his shorefront acreage was a direct result of wasteful delivery and irrigation practices. The state Department of Water Resources (DWR) investigated, agreed, and referred the case to the State Water Resources Control Board (SWRCB). That same year (1984) the SWRCB issued Decision 1600, which called on IID to conserve water by improving its distribution practices. Among the potential remedies if IID did not take action were loss of some of IID's water rights and intervention by the SWRCB in management of the district. IID claimed that the SWRCB lacked jurisdiction in this matter, but the claim did not hold up in court and on appeal. IID was forced to find least-cost ways to refurbish, modernize, and build new infrastructure and to alter management and irrigation practices in order to conserve water.

A tangible effect of IID's declining political power in federal and state affairs had been felt in the 1980 congressional redistricting. Imperial County had been joined with urban, populous, and growing San Diego County, reducing the district's influence both in Washington, D.C., and in Sacramento, the state capital. Some long-established, conservative farm families in the Imperial Valley did not take stock of the political realignment and did not take seriously the threat implicit in Decision 1600, but others did, and started taking action.

Meanwhile, farther west, along the coast, MWD had suffered a stunning defeat in its efforts to secure additional water supplies from northern California. A Peripheral Canal had long been planned to swing Sacramento River water (flowing south out of the Trinity Range and the northern Sierra Nevada) in a loop around the Sacramento–San Joaquin Delta. On the southern side of the Delta, the canal would hook up with the California Aqueduct and continue flowing south to MWD and other users in the southern part of the state. The canal was designed to prevent much of the Sacramento River's freshwater from flowing through the Delta and into San Francisco Bay so that the State Water Project could use it to deliver MWD's total contractual allotment of 2.01 maf/year.[8] In 1982, the canal proposal was put to a statewide referendum. It was resoundingly defeated.[9]

MWD interpreted the vote to mean that new north–south supplies would not be forthcoming until MWD could show that it had undertaken local conservation practices and made other water-supply arrangements not involving major infrastructure projects. MWD therefore refocused its search for new supplies from the north to the east. By 1984, IID and MWD were at the negotiating table.

Ill-Fated First Attempt at a Conservation/Transfer Agreement

IID hired Parsons Engineering[10] to determine just how much water it could realistically conserve. Parsons determined that the district had already saved roughly 138,000 af/year from previous conservation efforts but could save much more by lining earthen canals with cement, adding new reservoirs, and replacing leaky wooden gates. However, the required techniques would be expensive. To pay for them, the district contemplated transferring for a flat fee as much as 250,000 af of conserved water to other users. MWD was an obvious candidate to receive the water because it had both an urgent need and the financial resources to pay.[11] In 1985, after eighteen months of meetings, IID and MWD negotiators announced a forty-year program in which MWD would pay $100/af annually for ten years in exchange for a forty-year IID commitment to transfer 100,000 af/year. Physically, the transfer would work as follows: IID would reduce its diversions from the Colorado River into the All-American Canal by the amount saved, and MWD would divert the same amount upstream into its Colorado River Aqueduct.

Hardly was the announcement out the irrigation district's door when the proposed deal ran into a roadblock of criticism from IID farmers and residents. It was ultimately rejected on a 3–2 vote by IID's board of directors. Among the reasons for the rejection was skepticism that IID could comply with Decision 1600 by transferring *previously conserved* water and that MWD would actually pay $100 million over ten years for actions IID had already taken. Further, without some form of price index, high inflation could reduce the real value of the annual payments. Moreover, the payments might not cover power-revenue losses (due to less water flowing through the turbines along the All-American Canal) and potential costs of damage to the Salton Sea caused by higher concentrations of salts and other pollutants in the reduced farm drainage. Farmers also may have feared that although their water tolls would pay for future conservation investments, IID would keep the revenues from water sales, providing no direct benefits to the farmers. Finally, there was a general concern that allowing MWD even limited access to IID water could lead eventually to loss of local control of the water resource (Smith and Vaughan 1987). Having failed before IID's board, this initial agreement was never presented to MWD's board.

IID board's failure to ratify the original agreement stopped contact between the two parties for roughly eighteen months, but it did not stop momentum at IID for some kind of conservation/transfer agreement. By 1986, IID had completed an environmental impact report (EIR) on the infrastructural improvements recommended by Parsons. It again contemplated the ultimate transfer of 250,000 af/year from nearly 500,000 af/year of conserved water. It described a transfer of 100,000 af/year as an "initial transfer" and a "first step" toward subsequent transfers to an unspecified buyer. MWD remained the obvious choice, but with the collapse of their agreement, the two sides had begun mustering legal arguments concerning who should have rights to any conserved water in the absence of an agreement. MWD argued that it, CVWD, and PVID, as junior appropriators, had the right to any water IID conserved. But state law had recently been amended to include conservation and subsequent transfer as a "beneficial use." IID used this to argue that no additional water would be available for junior appropriators solely as a result of conservation practices. IID asserted that it had the right to transfer conserved water to any other party, be it MWD, the San Diego County Water Authority, or the Kern County Water Agency.

MWD's general manager, Carl Boronkay, in his comments on the draft

EIR, took the opportunity to lay out MWD's principles of negotiation for a conservation/transfer agreement (MWD 1986, p. 2):

> The initially proposed project should consist of making the first 100,000 acre-feet per year of newly conserved water available to Metropolitan. . . . Metropolitan is willing to fund new conservation measures, not conservation measures which are already in place.

Boronkay later added the proviso that the cost of new conservation projects must be "economically viable." MWD stood by these principles throughout subsequent negotiations and execution of the agreement.

Meanwhile, following the IID board's rejection of the IID–MWD agreement, the SWRCB began a series of detailed hearings on IID's compliance with Decision 1600. It began considering its own set of conservation measures for IID.

In February 1988, IID offered to provide 100,000 af/year for thirty-five years at a cost of $17.5 million per year, a 75 percent increase over the 1985 price. MWD rejected the offer, instead turning its focus to the SWRCB to press for imposition of a conservation program. It also began lobbying in Washington for congressional support for a concrete-lining project along the All-American Canal. With Congress's blessing, MWD would pay for the lining project and derive the benefits of the conserved water, bypassing IID altogether. This latter effort set off alarms not just in Imperial County, which wanted no outside control over any aspect of the region's water infrastructure, but also in Mexico, where seepage from the canal is captured and used to irrigate vast stretches of borderland farms.

Two events later in 1988 pushed IID and MWD back to the table. The first was the SWRCB's Order 88-20, issued in September. It required IID to submit by January 1, 1989, a specific plan to conserve at least 100,000 af/year by January 1, 1994. The penalty for noncompliance would probably be intervention by the SWRCB in management of IID's water systems.[12] Intervention by any outside party in IID water management was the region's worst-case scenario. It is fair to say that this order, like a shot across the bow, reinvigorated IID with the sense of urgency brought about by Decision 1600.

The second event of 1988 was a reshuffling of IID's board of directors during the regular 1988 elections, the result of which was that the board now had enough votes to pass a water-transfer agreement.

Fast-Track Negotiations Reach an Agreement

When the parties returned to the table, MWD wanted to acquire firm rights to water that would make up for the possible loss of some of the district's Colorado River flows. However, the district would pay only for new conservation measures that would allow it to capture additional flows as a junior appropriator. This position posed risks for MWD. It would have to pay IID up front for future conservation performance over which it would have indirect control at best. IID, on the other hand, faced risks no matter what it did. If it refused to comply with the SWRCB's order, it could lose some of its water rights, control of its daily operations, or both. If it did comply, its conservation/transfer program could launch an ongoing erosion of its water rights for the benefit of urban consumers.

Therefore, IID sought assurances that it would regain its rights to either use or resell its conserved water at some time in the future. For its part, MWD needed to ensure that its investment in IID's infrastructure would increase its water allocation on a consistent basis. The negotiations were further constrained by IID's need to comply with the SWRCB's January 1, 1989, deadline for a conservation plan. The technical nature of the talks—identifying potential conservation projects and estimating costs and water savings—brought engineers together from the two sides. Their comfort level rose as more opportunities were confirmed for actual savings of "wet" water, as opposed to "paper" water. Moreover, the talks also benefited from both sides' familiarity with the basic mechanics of the transfer, based on past negotiations.

Second Agreement, Reactions, and Revisions

Less than two months after IID received its copy of Order 88-20, the district had an agreement with MWD. Titled "Agreement for the Implementation of a Water Conservation Program and Use of Conserved Water," it is forty-five pages long and includes sixteen articles and five appendixes. Both boards voted their approval of the agreement before the end of the year, permitting IID to comply with the SWRCB's order.

The parties agreed to a two-part program. The first part was to last five years and consisted of implementing sixteen separate conservation measures, all specified in the agreement. In total, the measures were expected to conserve 100,000 af/year. During this period, MWD agreed to reimburse IID for its capital expenditures, projected to be $98 million, as well as operating expenses once the measures were installed and working. MWD also agreed to make five annual payments of $4.6 million to cover "indirect" costs. In the sixth year, with the conservation measures completed, MWD's right to conserved water would begin. For the following thirty-five years, MWD would have the right to divert 100,000 af/year into its Colorado River Aqueduct on the basis of the conservation and transfer of 100,000 af of IID's third priority. MWD would continue to pay estimated annual expenses of $2.6 million in constant 1988 dollars.

With the agreement, IID had hoped to secure both ownership of its water rights and control of its infrastructure. It got the latter. The agreement clearly and repeatedly states that IID is in charge of feasibility planning, cost estimation, final designs, plans, construction management and inspection services, and environmental documentation.[13] The agreement affirms that IID is in charge of all ongoing maintenance and provision of remedies for loss or destruction of any project.[14]

The water rights issue, however, was left unsettled for IID. The final recital at the beginning of the agreement reads, in part:

> IID and MWD recognize that they have differences of opinion over various legal questions relating to the transfer of certain water and the entitlement of junior priorities to certain water, but each wish to go forward with the Agreement. . . .[15]

Indeed, the two sides began to both lay groundwork for and erect barriers to future legal wrangling over disposition of the rights.[16] For example, IID obtained MWD's agreement not to argue that it has a post-agreement-period right to the conserved water based on the fact that it paid for the conservation projects.

One definitive agreement on the nature of the transfer is that the conserved water "shall at all times retain its 'third priority' status. . . ."[17] In other words, the agreement is to be construed as IID's conserving and transferring water from its third priority, not as IID's either losing part of its right or choosing not to consume its full right in order to make additional water available for MWD's fourth and fifth priorities. MWD agreed to this because CVWD and PVID (the latter for its mesa lands) stood between IID and MWD on the priority list, so there could be no guarantee that MWD would get the water it was paying IID to conserve, especially in dry years, if IID were simply to "leave it in the river."

MWD did, however, gain contractual assurances that (1) during the first part of the program it would not be overcharged for the actual conservation measures and (2) during the second part, it would have assured use of 100,000 af/year of IID's water. MWD's assurances on the first point arose from its membership on the Program Coordinating Committee (PCC), which was created to oversee the program in its entirety.[18] The PCC has three members: one from each agency and a mutually agreed-on third member who serves as chairperson. Agreements are made by majority vote. MWD also can take the retroactive steps of auditing expenditures on the program and challenging them within one year of their submission for reimbursement.[19] If the PCC cannot resolve an accounting dispute or any other type of dispute, the two agencies' general managers have thirty days to try to resolve it. Failing that, a process for binding arbitration is specified.[20] These provisions keep MWD at arm's length from all physical activities in the Imperial Valley but give it at least some oversight capabilities.

Also incorporated in the agreement are cost estimates of all sixteen conservation projects and an incentive clause stating that if the overall cost of the program is less than the estimated total, MWD will pay 25 percent of the savings into IID's discretionary Indirect Cost Fund. The agreement includes such flexibility-enhancing elements as feedback mechanisms for project implementation and a right to substitute more effective conservation measures for less effective ones.

With respect to long-term assurances, there are few. IID agrees that if the water conserved ultimately does not total 100,000 af/year, it will undertake additional conservation measures until that figure is achieved, to be paid for by MWD.[21] In dry years, however, IID retains its right to take water for itself first. That is, if the Bureau of Reclamation is unable to supply IID's full allotment, IID has priority to consume as much as the Bureau delivers prior to making any available for MWD.[22] For every dry year in which MWD does not receive water from IID, an additional two years are added to the agreement. Still, the parties concur that with respect to future default by IID,

> as to MWD the remedy of damages is inadequate and unavailable to enforce this Agreement.[23]

The parties therefore pledge to continue their performance under the agreement until a judicial resolution has occurred.

What is probably the agreement's most important security feature is not even mentioned in the agreement: the SWRCB's Order 88-20, which mandates conservation and threatens intervention. If IID fails to conserve and transfer water in a manner consistent with the SWRCB's order, it risks losing both local control and a portion of its water right. Thus, MWD gained weak contractual assurances of long-term performance but could rely on the external circumstance of the credible threat that IID might lose a portion of its water rights if it breached the agreement.

National Acclaim and Local Opposition

What two years previously had been dubbed by observers a failure with stymied negotiations (Smith and Vaughan 1987, p. 72) was now hailed as an outstanding example of a mechanism to meet western water needs. Secretary of the Interior Donald Hodel said that the agreement reflected "a new evolution in water resource thinking, an evolution of which the Department of the Interior is entirely supportive" (*Calexico Chronicle* 1989). The Environmental Defense Fund published an article coauthored by the two agencies' general managers in which they called the program an "encouraging example" that should "serve as a model" for other water agencies (Boronkay and Shreves 1989). Awards started rolling in, including a 1989 Water Conservation Award from the Bureau of Reclamation, a 1990 National Water and Energy Conservation Award from the Irrigation Association, and, in May 1991, the Clair A. Hill Agency Award for Excellence, sponsored by the Association of California Water Agencies. To this day, the IID–MWD agreement remains the highest-profile example of a successful water-market transaction in California, perhaps in all of the West.

Locally, however, opposition to the program was growing. One source of opposition was the Coachella Valley Water District, located at the northern end of the Salton Sea. Representatives of CVWD had two primary concerns. First, they believed that the district had the right to any water conserved by IID because it was next in line within the third priority. Second, they thought that overall, the agreement had the potential to reduce CVWD's right to use Colorado River water. By February 1989, CVWD had filed a federal lawsuit seeking to nullify the agreement, and soon the district was engaged in negotiations with IID and MWD.

Within IID's service territory, opposition among traditional elements of the farming sector, as well as within agriculture-related sectors, was growing. On Christmas Day 1988, just three days after MWD's board would approve the agreement, Rev. Bobby Adams placed an advertisement in the local newspaper, the *Imperial Valley Press* (Adams 1988). The advertisement included a Christmas blessing, underneath which was a picture of professional wrestler Hulk Hogan sneering and clenching his fists, captioned with the words "Please sign the recall petition against those on the I.I.D.

Board, who deal away our precious water!!!" A news article announcing the agreement reported:

> The deal was put together after a tumultuous year that left IID in a weakened bargaining position and in essence forced an agreement to be made. (Morris 1989)

Opponents called for a public vote. They framed the issue as a question of whether agreeing to this program made inevitable a subsequent program, which could include taking farmland out of production. Farmers with large landholdings who were satisfied with the agreement, as well as lead IID negotiator Don Cox, opposed the vote. It ultimately was not held. In January, Cox defended the agreement in an editorial (Cox 1989):

> With its back to the wall and having fired the last of its legal torpedoes at the State Water Control Board and achieving no hits; it became obvious that [IID] needed to make a deal with [MWD].

In February, the first conservation-related construction was under way. MWD would be sending its first check at the end of the year.

In May, CVWD proposed a settlement to its lawsuit whereby MWD would agree to make water available to CVWD if the agreement reduced its water supplies. Negotiators from IID and MWD tentatively agreed to this. The proposed settlement provided a final opportunity to review and reconsider the overall program. If IID went to court with CVWD and lost, there would be no agreement; if it settled with CVWD in a way unacceptable to MWD, that also would result in no agreement.

By late July, public opinion about the program was still wavering. Bill Condit, an IID board member who was not a farmer, revealed that he had tried to get MWD to pay for covering the open canals that passed through the valley's towns. The conservation rationale was that a pipe would cause less evaporation than an open canal. This was an emotional issue for town dwellers because there had recently been two drownings in the open canals, one involving a child. Condit explained that from MWD's perspective, this was "an expensive way to conserve" water, and so the project was not adopted. Sentiment against the agreement did not abate as summer gave way to autumn. On October 5, a "Voice of the People" editorial in the *Imperial Valley Press* called the agreement "just another version of the first contract that we stopped in 1985" and warned against both loss of control of the irrigation infrastructure and escalating prices for irrigation water (Severns 1989).

Others noted, however, the benefits that would soon flow into the valley. First, MWD's creation of a $23 million fund for IID to cover "indirect costs" helped reduce concerns. Originally meant to cover hydropower-production losses and potential environmental liabilities arising from more concentrated effluent streams entering the Salton Sea, the fund ensured that money would be available to cover other injuries associated with the conservation program or even such public benefits as covering canals that ran through valley towns. Second, in October, the SWRCB's chairman, W. Donald Maughan, wrote a letter to IID that reiterated the board's preference for local solutions "to avoid the need for more direct interventions by the State Board or courts." Maughan continued, "Timely implementation of the program which the District has developed, however, is essential to minimizing State Board or judicial involvement in IID operations in the future" (*Imperial Valley Press* 1989, p. A-6). This bolstered the claims of many supporters that a no-agreement scenario could be far worse than what they had. Finally, the first MWD payment would be arriving in January 1990, provided final agreements satisfying CVWD and the other parties could be reached. The payment would be approximately $18 million for direct expenditures plus an additional $4.6 million for the Indirect Cost Fund.

Subsequent Agreements

Two additional agreements were finalized and signed in time to save the program. The first, called the Approval Agreement, was intended primarily to satisfy CVWD's desire for protection from injury resulting from the program and to further satisfy IID farmers that local control would not be compromised. In essence, the agreement stipulates that rather than conserving and transferring a portion of its third priority, IID is not consuming all of its third priority and is leaving the balance in the river. CVWD and PVID, which share the third priority, agree not to take the conserved water left in the river except in certain years. This water is then made available to MWD's fourth and fifth priorities.

The reason for this switch is that under the new rights regime, CVWD again has priority over MWD. In dry years when the river is "on order" and the Bureau of Reclamation cannot satisfy all appropriation requests, CVWD has the right to take as much as 50 percent of IID's conserved water, according to a proportional strategy elaborated in the Approval Agreement.[24] This scenario was not expected to occur for several years, pending Arizona's more complete utilization of its water from the Central Arizona Project.

The agreement also creates a Water Conservation Measurement Committee to oversee efforts to estimate actual amounts saved and to make a final determination of the amounts saved. The committee makes decisions only with a unanimous vote. In the event unanimity is not achieved, a dispute-resolution procedure involving arbitration is to be followed. MWD agreed to pay for consulting services to support this committee, and IID agreed to cover its general expenses.

Finally, the agreement gave some relief to Imperial Valley growers, which served to "seal the deal." Three of the original sixteen projects enumerated in the agreement were dropped, and two new projects were added. Of those dropped, one was discarded; the others formed a new, separate list called the Augmentation Program. The discarded project would have increased surveillance efforts by IID personnel for evidence of overwatering by farmers. Observation of excessive tailwater has historically resulted in triple charges for that day's water, but the rule had not been heavily enforced.[25] Some tailwater percolates into the tile drainage system that underlies much of the valley and then flows into drainage canals destined for the Salton Sea; other tailwater flows directly into open drains and then to the sea. This measure, which would have added eighteen additional IID staff members to survey tailwater in the district, was expected to conserve 5,000 af/year. It was the only measure on the list that involved penalties for farmers, who were relieved to see it dropped.

The two measures transferred to the new Augmentation Program were (1) the building of a 340-af reservoir and pumping facility to conserve water that otherwise would have been discharged into a drain and (2) the lining of a two-mile section of a major canal, with total expected savings of 6,110 af/year. IID had already undertaken these two projects, financing them with loans from the state. IID agreed to repay those loans itself and make the water available to MWD for free in partial compensation for the new risk to MWD posed by CVWD's taking of part of the conserved water in dry years.

These two new measures, when added to the original list, increase the overall expected savings to 106,110 af/year, all of which will normally be available to MWD. The Augmentation Program was one of the last elements added to the agreements, offered by IID to secure final approval and get the contracts signed.

With the finalizing of the additional agreements, CVWD withdrew its lawsuit, the SWRCB relaxed its scrutiny, local groups accepted the program, and the implementation process became the program's focal point. A total of five years of negotiations, three agreements, and a side letter from MWD related to water banking had brought the parties to this point.

Executing the Agreements

Two strengths of the program have turned out to be its dispute-resolution systems and its implementation flexibility. Of the three main issues that have arisen, all have been handled within the structure of the agreement. The first issue concerned the price of dirt.

In February 1991, an issue was raised that IID was using its farmers' dirt for flood control and other purposes related to the program without compensating the farmers. By April, IID's board had agreed to negotiate with farmers a value for the dirt, which would be passed along to MWD. When the first bill came to more than $1 million, MWD's general manager, Carl Boronkay, sent a letter to IID stating that this amount was too much to pay for dirt. In response to the letter, IID's board revised its dirt compensation program to one concentrating on farmers' claims that the dirt was unfairly taken, and the issue in general died down. One additional sticking point, however, was that IID paid $90,000 in a settlement for dirt removal in which a brother of an IID board member served as negotiator. IID defended the negotiation as legitimate, and the district's general manager, Charles Shreves, agreed to write a letter to Boronkay to that effect. This dispute, involving both cost and procedural issues, was handled effectively through direct communication between the two districts' general managers, exactly as designed.

The second issue involved the feedback, evaluation, and course-change mechanisms that guided selection of conservation measures. Because of the SWRCB's time constraint in finalizing the agreement, a number of the conservation projects listed were still in a "prefeasibility" stage when included in the program. One example of contractual flexibility proving to be beneficial concerned plans for canal lining. The agreement envisioned the lining of 265 miles of earthen canals, which would conserve 29,150 af/year. Without the lining, that water would either percolate into the aquifer or seep out and form standing pools near the canals. This was the first measure undertaken, since IID had considerable in-house expertise. After eighteen months of work and completion of 110 miles of lining, the verification process "caught up with" the construction efforts. IID engineers found that, in part due to variable flow regimes, canal lining would conserve substantially less water than originally anticipated. In response to these findings, the lining program was scaled back and an additional test was introduced to prequalify earthen canals before any concrete was laid.

Offsetting the discouraging results of canal-lining efforts were the positive results achieved by construction of the Plum-Oasis Lateral Interceptor, the first interceptor built in IID. A lateral usually is a straight, narrow canal that is connected on one end to the main canal, delivers water directly to farmers' fields along its course, and then dead-ends. If too much water moves from the main canal into the lateral or if farmers along the way do not open their headgates and take what is there, the extra water keeps moving down the lateral and could spill over the far end. A lateral interceptor captures that spilled water. It consists of an open canal that collects and transports both operational discharge water (water spilled prior to delivery to a farmer's field) and overflow of water delivered to farmland. With the interceptor canal, the spilled water bypasses the drainage system and instead is transported to a reservoir, from which it can be recaptured and reused in another part of the distribution system. The Plum-Oasis project not only conserved water but also increased farmers' flexibility as to when they could shut off a delivery of water without loss to the system. At least two additional interceptor projects were added to the program. The flexibility of allowing the Program Coordinating Committee to review progress and results and make course corrections consistent with economic achievement of the conservation goal is considered by MWD's PCC member Arnold Dimmitt to be "instrumental to [the program's] success to-date" (Dimmitt 1994).

Perhaps the biggest test of the program's flexibility involved the environmental challenges that arose because of the program's potential for increasing the concentration of agricultural effluents in runoff bound for the Salton Sea. Longtime IID participants in the negotiations believe environmentalists' strategy in supporting the conservation/transfer plan was to sacrifice the Salton Sea in order to save the Sacramento–San Joaquin Delta. As the strategy went, if MWD could get the extra water it needed from the Colorado River, it would demand less from north–south conveyance, thus easing environmental pressure on the Delta. These IID parties feared, however, that once conservation measures were taken and measurable adverse effects were seen in the sea, local environmental pressure would be renewed. By mid-1990, this was happening. A group called the Salton Sea Task Force was raising concerns about the effects of reduced inflow on the sea's salinity level and was preparing a letter to the SWRCB. A U.S. Fish and Wildlife Service manager at the Salton Sea National Wildlife Refuge expressed concern that the sea could become a "death trap" for birds and other aquatic life if salt and selenium concentrations from agricultural drains increased. The draft letter raised the possibility that "parties benefiting from the conserved water should be legally responsible for the mitigation of these impacts" (Rice 1990b).

The SWRCB's Regional Water Quality Control Board threatened to sue IID if it did not prepare supplemental environmental impact reports based on the new data concerning high concentrations of selenium and other toxins in its drainage water. The regional board would consider cumulative effects of projects already undertaken as well as new plans for two lateral interceptors and an additional project that had not yet been specified. Until the reports were complete, no work would be performed on projects that might directly affect the concentration of effluents in agricultural drains.

The IID–MWD conservation program survived this temporary curtailment. For two years, no construction took place on roughly $30 million worth of conservation projects, but eventually they were back on line, with expected completion in 1997 instead of 1994. The SWRCB approved this extension of its original deadline, as did MWD. The agreement recognizes that when failure to carry out the agreement is due to an "uncontrollable force," due diligence to remove the inability is all that is needed to maintain the agreement. As a sign of commitment to the agreements, MWD made its five annual indirect cost payments of $4.6 million on schedule.

Seven years after the original signing of the Water Conservation Agreement, it was still in place and most of the projects had been completed. As water-conservation measures have been completed, MWD has diverted water equivalent to the incremental savings. In 1990, that amount was 6,110 af (the augmentation-plan savings); in 1994, it had risen to 72,870 af, and in 1997, to 97,740 af. The thirty-five-year period began in 1998, when the full contracted savings were achieved and taken by MWD.

Although some participants in the process were concerned that CVWD might make one more attempt to garner concessions from the two parties before commencement of the thirty-five-year delivery period, there appears to be no fatal hindrance to MWD taking the conserved water if it cannot otherwise satisfy its fourth and fifth priorities.

Effects of the Agreements

Many of the consequences of the transfer have already been described, but they are summarized here. They fall into the categories of local and non-local effects.

In terms of local effects, the water conservation measures have reduced flooding for owners of land littoral to the Salton Sea. This was the problem that prompted farmer John Elmore to file his suit in 1984. Groundwater elevations have also been lowered in IID. This has been of benefit to farmers because a high groundwater elevation can concentrate salts in the root zones of plants, impeding their growth.

IID received a $100 million infrastructure upgrade. It now has a state-of-the-art Supervisory Control and Data Acquisition System in a new building. Microwave communication devices link remote sensors that automatically adjust canal gates to achieve targeted flow rates and elevations throughout the system. IID has replaced leaky, decades-old wooden gates with aluminum gates and built its first lateral interceptor system. It currently is able to satisfy all previous normal-year needs for water using 100,000 af/year less than before. IID has made the effort; MWD has footed the bill.

There have been positive effects for the labor force in construction of conservation measures and perhaps less positive social adjustments for the *zanjeros,* or "ditch-riders," who were responsible for opening the gates that provided water to the farmers' fields. The new, more highly automated system changes their role, reducing opportunities for the brief, friendly conversations with farmers that had always been a part of valley farming.

Another local effect involves the program to cover canals that pass through valley towns. IID's board agreed to dedicate half of the interest income from the Indirect Cost Fund to a matching-fund program for covering the canals. Both local and county officials had been asking for this program for some time.

To the minds of many people in the Imperial Valley, what the program helped IID avoid is as important as what it has done. Threats of seizure of irrigation-infrastructure management or of water rights receded. Fears of losing local control of resources to the urban southern coast also were reduced.

Nonlocal effects include provision of the long-sought example of a successful water-market transaction. The importance of the transfer is summed up in a July 21, 1991, article in the *San Diego Union.* Titled "State Starts Water-System Reform: Five-Year Drought Has Demonstrated Marketing Is the Way to Go," it spends nearly all of its roughly 600 words in making the case for water marketing but provides only one example: "A water-transfer agreement made by . . . [MWD], which supplies San Diego's water, and . . . [IID], serving agriculture east of here, demonstrated that water markets can work" (Kaplan 1991).

Report Card: How Well Did the Agreements Mitigate Water-Transfer Risks?

For MWD, the two primary contracting risks involved seeking guarantees that it would receive the water it had paid for and overseeing IID's water-conservation performance. A third risk involved securing the conserved water from use by intervening priority holders.

MWD's first risk amounted to paying money today for future performance. To mitigate the risk, Carl Boronkay could have asked for an agreement that matched payments with water deliveries over time, but IID needed capital in advance of water deliveries to pay for the expensive engineering projects that would yield conserved water. Alternatively, MWD could have attempted to build long-term performance guarantees into the agreement. It chose the latter path and followed it in two ways.

First, IID agreed to continue to make water available to MWD according to the terms of the agreement until any future dispute is resolved. As an open-ended guarantee to supply water throughout a dispute, it is a solid security feature. It does not, however, address the underlying risk of a dispute arising. Second, MWD's deliveries are secured by the regulatory or judicial risk IID faces if it stops conserving water. The SWRCB is satisfied with the IID–MWD agreement because it means that IID is wasting less water. If the conservation program is stopped (and wasteful practices are resumed by IID), the SWRCB or a court could intervene and seize IID's water rights, management, or both. This potential outcome is so undesirable to IID that it provides a strong incentive for the district to live up to its contractual agreements. The threat of intervention provides MWD with only partial security, however, since it does not stop IID from transferring conserved water to another party. Possible other parties include the cities of San Diego and Las Vegas or Kern County.

MWD's second risk involved the extent to which it could exert control over both IID's performance in carrying out conservation measures and the cost of performance. Among the options available to MWD were active participation in conservation measures, heavy on-site scrutiny of IID performance, and hands-off ex post facto monitoring of performance.

The extreme sensitivity of IID's farming community to urban intervention in their agricultural activities meant that MWD was effectively barred from active participation in or oversight of projects. Instead, a less obtrusive oversight role was necessary. This was accomplished primarily through MWD's participation on the Program Coordinating Committee.

MWD also had audit rights and back-up methods for dispute resolution, including manager-to-manager negotiations and binding arbitration.

Further, the agreement includes an incentive clause: if IID's actual conservation costs are lower than those listed in the agreement, it can keep 25 percent of the savings. IID does not regard this as a major incentive, however, primarily because any savings are to be deposited into the Indirect Cost Fund instead of an IID general fund.

As described earlier, the dispute-resolution apparatus worked effectively when MWD believed it was being overcharged for dirt used in construction projects. The two general managers were able to resolve their differences, and the overall agreement proceeded.

With respect to risks regarding the Bureau of Reclamation's management and delivery of Colorado River water to MWD's aqueduct, this has not been one of MWD's major concerns. The Bureau of Reclamation oversees flows in the Colorado's upper and lower basins and ensures that appropriate amounts are delivered to the aqueduct intake at Lake Havasu. The water-supply risk to MWD is well defined and related to its fourth and fifth priorities for California's share of Colorado River water, not to management of flow or delivery of the water.

The third risk involves the potential that the other districts sharing the third-priority right with IID could intervene and take the conserved water. This problem could not be erased by contracting; instead, it was bounded and compensated. A detailed formula restricts CVWD to half of IID's conserved water in dry years. To make up for this loss, IID increased the total amount of conserved water available to MWD annually and agreed to extend the agreement in relationship to the frequency with which CVWD takes conserved water.

In comparison with MWD's historical rights, this arrangement is equally insecure in dry years, when MWD most needs secure rights. Guaranteed delivery in wet years is scant compensation for curtailed delivery in dry years, but at least it will allow MWD to fill local reservoirs or aquifers in anticipation of future needs.

In sum, new risks and security features were introduced into the IID–MWD agreement that did not exist in MWD's original Colorado River import agreements with the Bureau of Reclamation. Some options that might have increased MWD's security, such as a pay-as-you-go arrangement or greater oversight capabilities, were not available because of the specific circumstances of the negotiations and the parties. Still, every concern expressed by MWD was secured to some extent through the contracting process.

Turning to IID, that district's primary contracting risks involved erosion of water rights and loss of control over existing operations. In terms of erosion of rights, IID faced risks in both no-agreement and agreement scenarios. The no-agreement risks included imminent SWRCB or judicial reduction of water rights by the amount considered to be wasted, loss of management control over the district to the SWRCB, which would then have imposed conservation measures, or both. IID could substantially reduce these risks either by entering into a conservation/transfer agreement that satisfied the SWRCB's Decision 88-20 or by simply self-financing water conservation.

With a transfer agreement, IID faced a similar risk of erosion of property rights, but for different reasons. Since water rights are secured over time by regular use, once IID stops using some of its water, it risks losing its rights to the water. IID originally tried to reduce this risk by characterizing the conserved water as still belonging to its third priority. A nonpermanent water transfer is now considered to be a beneficial use of water; a transferor is likely to get the water back at the end of the transfer period. This approach was foiled, however, by CVWD, which successfully argued that its own water rights would be harmed by such an arrangement. Consequently, IID was forced simply to leave its conserved water in the river for use by a subsequent priority holder, MWD. IID then sought contractual guarantees that at the conclusion of the agreement (some four decades later), it could reclaim its property right.

Toward this end, the agreement specifies that MWD could claim no postagreement rights to the conserved water on the basis of its having paid for the infrastructure.[26] This is a rather weak security feature because it does not rule out other claims to the water that MWD (or others) might someday stake, including a claim that thirty-five years of nonuse amounts to abandonment of a water right.

Compared with IID's secure, federally recognized "perfected" rights to 2.6 maf of water per year, the risk of ultimate loss of the conserved 106,000 af/year was very high. But in relation to the regulatory threat posed by the SWRCB's Order 88-20, the financial consideration offered by MWD, and the modest contractual protection against a future MWD claim, the agreement represents a resourceful outcome to a severe challenge.

IID faced possible loss of control over its operations to the SWRCB in a no-agreement scenario and possible temporary loss of control during the implementation phase of conservation projects. For political and historical reasons, even temporary loss of control was a major concern to Imperial Valley farmers.

As a result, the agreement is very clear in putting IID in charge of all aspects of practical implementation of the conservation projects, from start to finish. MWD has no role beyond membership on the Program Coordinating Committee and ex post facto auditing and review. IID has designed, managed, and executed all of the conservation projects. This arrangement calmed local concerns enough that IID board members were able to ratify the agreement. IID was therefore able to maintain its historical role as water manager for the Imperial Valley.

The existence of one such transaction suggests that additional agreements are possible. Yet no subsequent agreements between these two parties have been signed, although a possible transfer between IID and the San Diego County Water Authority was in an advanced stage of development at the time of this writing. Negotiations between IID and MWD have examined a range of measures, from additional conservation to temporary land fallowing to major canal-lining projects. Only the one set of 1989 agreements that enabled IID to meet its regulatory requirements in a least-cost manner has been signed and executed. Without the threat of sanctions hanging over IID, an agreement might never have been signed.

In conclusion, although the IID–MWD agreement is hailed as an important example of a water-market transaction, in its details one finds a complex arrangement involving substantial risks for both sides and numerous security features. The complexity of the agreement is overlain by the conceptual simplicity of the arrangement. Ultimately, no water *rights* transfer ever occurred. One senior appropriator is simply reducing its use of the river and a subsequent appropriator's use is increasing. For all that the IID–MWD agreement does portend in terms of potential for rural–urban cooperation, it does not appear to portend a statewide water market with small-scale participants, small-scale transfers, and numerous transactions.

Notes

1. In fact, five hydropower plants along the fifty-mile All-American Canal provide additional income to IID and CVWD.
2. *Arizona v. California,* 373 U.S. 546, 83 S. Ct. 1468, 10 L. Ed. 2d 542.
3. *National Audubon Society v. Superior Court of Alpine County,* 33 Cal. 3d 419, 658 P.2d 709, 189 Cal. Rptr. 346 (1983).
4. The Yuma Project is a series of water-rights contracts between the federal government and farmers on Indian reservations. Most of the reservations are located along the Colorado River near the Mexican border. Tribes with Colorado River rights are known locally as "wet tribes" because of the strength and high priority of their water rights.

5. The city of San Diego originally shared the fifth priority with MWD, holding rights to 112,000 of the priority's 662,000 af, but during World War II it ceded these rights to MWD. San Diego lies in MWD's service territory.

6. Contract No. 7-07-30-W0171, dated September 9, 1987.

7. *Elmore v. Imperial Irrigation District,* 159 Cal. App. 3d 185 (1984).

8. The most water delivered to MWD from the State Water Project was 1.4 maf in 1990, and the least delivered was 381,000 af in 1991 (MWD 1994, 5-3).

9. The Peripheral Canal proposal was back again as part of the CALFED Bay-Delta Program in the late 1990s. The proposed canal's new name was the Cross-Delta Facility.

10. Parsons Engineering is now called Parsons Infrastructure and Technology Group, Inc.

11. The original suggestion that MWD cover the cost of conservation efforts in IID in exchange for the saved water is broadly credited to Tom Graff of the Environmental Defense Fund (EDF). This idea was published in a 1983 EDF report titled "Trading Conservation Investments for Water" (Stavins 1983).

12. The order concluded that the SWRCB reserved jurisdiction to review and monitor the plan and its implementation and to take "other such action as may be appropriate to ensure that the requirements of Article X, Section 2 of the California Constitution are met."

13. See Section 1.5, Article I, "Agreement for the Implementation of a Water Conservation Program and Use of Conserved Water," between the Imperial Irrigation District and the Metropolitan Water District of Southern California.

14. Ibid., Section 1.8.

15. Ibid., Recital G.

16. Ibid., Article VI.

17. Ibid., Section 6.2(f).

18. Ibid., Section 1.2.

19. Ibid., Section 4.7.

20. Ibid., Article VIII.

21. Ibid., Section 3.3.

22. Presumably, MWD's share would then precede CVWD's share, since CVWD agreed in 1934 to give IID first priority to waters delivered via the All-American Canal.

23. Agreement, Section 5.2(a), which deals with the possibility of default by IID in the form of a refusal to make water available.

24. See Section 3.2 of the Approval Agreement (MWD 1989).

25. The surveillance provision was originally part of the agreement because of the possibility that the SWRCB could target IID's estimated 400,000 af/year of tailwater as an example of not putting water to reasonable use.

26. Agreement, Section 6.2(d).

CHAPTER 6

Timing Was Everything: The Palo Verde Irrigation District–Metropolitan Water District "Two-Year Test Land Fallowing"

> Soon after Chuckawalla Valley State Prison opened in 1988, complaints were registered about the prison's standard-issue metal outdoor benches. In the region's 120 degree Fahrenheit daytime summer heat, they were burning the skin of any inmate who tried to sit down.
>
> —heard in Blythe, California, 1995

Three years after announcing its water-transfer agreement with the Imperial Irrigation District (IID), the Metropolitan Water District of Southern California (MWD) announced that it had closed another deal. This time, the water transferor was the Palo Verde Irrigation District (PVID), located in the southeastern corner of California along the Colorado River. Instead of the irrigation district conserving water, as was the case with IID, the valley's farmers individually would fallow portions of their fields. Again, the unused water would be left in the Colorado River for MWD, a junior appropriator.

This chapter is similar in format to the previous one, offering a broad story of a water transfer that hinges on the transfer's contractual agreement. Background sections describe the constraints and opportunities faced by MWD, PVID, and PVID farmers. There follows a chronicle of the actual negotiations, the signing of the deal, and efforts to implement the agreement. The elements of the agreement that minimized the parties' transaction risks are then described. To the extent that these elements can be recreated elsewhere, the possibility of using a market mechanism for long-term transfers increases.

The Situation in the Palo Verde Valley

These background comments supplement those at the beginning of chapter 5. This section introduces the city of Blythe, located in the Palo Verde Valley, and the region's labor and environmental situations.

City of Blythe

Blythe is the Palo Verde Valley's primary commercial and population center. Median income grew by 60 percent in the 1980s, reaching $21,540 in 1990, while population grew by 26 percent, to 8,500. In the period 1990–1995, the town's population swelled by another 60 percent to more than 13,000 people, constituting roughly 70 percent of the total population of the Palo Verde Valley. Other unincorporated residential and business areas in the region include East Blythe, Palo Verde, Ehrenberg, Quartsize, and Ripley, all much smaller than Blythe.

Since the mid-1980s, Blythe has made an effort to diversify its economy in order to decrease its dependence on agriculture. Farming used to be the town's main source of income, but by 1994 it accounted for only about 40 percent of the total. The valley's farmers have reduced their commercial dealings with the town by becoming more self-sufficient—for example, using field-based packing methods instead of shipping produce to Blythe for packing. In 1993, only three of the valley's top seventeen employers were related to agriculture: a food processor ranked sixth, PVID eighth, and a feed processor tenth. In recent years, Blythe pursued and won contracts for the Chuckawalla Valley State Prison and Ironwood State Prison, both located about seventeen miles west of town. The two facilities currently employ about 1,800 people at an average annual salary higher than $37,000.

Agriculture in the Valley

In general, PVID has a high percentage of landowner-farmers; those landowners who lease their property for others to farm usually are retired farmers who still reside in the valley. The major exception to this pattern is an electric utility company that bought land to secure water rights for a power plant that has not yet been built.

The valley's agricultural labor force is divided between local full- and part-time employees and nonlocal seasonal laborers. Farmers typically hire a small full-time staff and supplement it with repeat temporary employees. These "regular" employees carry out such tasks as weeding and thinning plants and installing sprinkler pipes. Farmers hire harvesters through the valley's half-dozen labor contractors. The United Farm Workers of America has a presence in the Palo Verde Valley, and the citrus farms, encompassing some 2,000 acres, are unionized. The valley's 450 unionized lemon pickers earn roughly $80 to $100 per day between August and February, compared with 1995 wages of $4.25 per hour for other agricultural labor. Historically, 3,000 to 4,000 laborers worked in the valley, but as vegetable production has declined and as labor-substituting technologies such as alfalfa balers have been adopted, labor requirements have also declined. According to state labor officials based in Blythe, during the December–March peak harvest period for lettuce, broccoli, and cabbage, 2,500 workers, about 800 of them migrants, are employed. During other seasons, the valley supports about 800 agricultural workers. About half of the agricultural labor force resides in the valley. A large number are bused daily from Mexico; others live in the Imperial Valley and Arizona.

Natural Environment

Three biological plant communities dominate this desert region. Creosote bush scrub communities are found in undeveloped areas. Riparian habitats along the Colorado River provide a home for saltbush, quailbush, arrow-weed, and cattails. The cultivated areas provide habitat for such bird species as egrets, herons, and ducks.

Negotiating the Agreement

This section first presents the history of negotiations between MWD and PVID and then explores the interests of the parties to the negotiations. A discussion of the two districts' preliminary price strategies follows, as well as a description of the negotiations themselves. The agreements reached are the subject of the section that follows.

Multiyear Negotiations

MWD first approached PVID in 1986. The original conception of the transaction was a long-term (thirty- to forty-year) dry-year option agreement. In other words, at MWD's request, in dry years farmers would fallow their land and let the water "flow down" from their first priority to MWD's fourth or fifth priority. MWD offered farmers $200 per acre to join the program, plus an additional $400 per acre for each year that MWD exercised its option and diverted the water. The amount of water per acre to be available for transfer was estimated at 4.6 acre-feet (af).[1] Negotiations stalled over long-term price-escalation clauses. PVID wanted to link diversion payments (during dry years when the option was exercised) to the Consumer Price Index, whereas MWD wanted to link prices to that of a specified basket of goods. No deal was struck.

A number of conditions brought MWD and PVID back to the negotiating table in late 1991. Whiteflies had attacked cotton, vegetables, and alfalfa throughout the valley. IID farmers estimated crop losses from whitefly damage at $130 million in 1991 and $100 million in 1992; this represented roughly one-fifth of IID's total agricultural revenues. An additional change was a third year of depressed prices for alfalfa, wheat, and cotton. Alfalfa prices were especially weak, and because other crop prices were down, the strategy of crop diversification had not succeeded in balancing farmers' incomes.

The valley had also experienced a steady decrease in vegetable production, in part due to the whitefly but also due to stringent state regulations for testing of pesticides and herbicides. A number of chemicals used in Arizona either had not been (and still are not) certified by the California Environmental Protection Agency or were more expensive in California because of the need to recover the cost of testing. The decline in vegetable production made farmers more dependent on alfalfa and other grains and therefore more vulnerable to weak grain prices.

By the early 1990s, MWD also had new motivations for negotiating. The state was experiencing a multiyear drought that made water originating in the Rocky Mountains all the more important to southern California. But just as MWD's demand for Colorado River water was growing, other users in the lower basin were increasing their demand. In 1990, due in part to completion of the Central Arizona Project and in part to drought conditions, the Colorado River was for the first time unable to meet the requirements of California users.

Contact was reinitiated in late 1991, and soon negotiations were under way. MWD's side was represented primarily by Bob Schempp, who had led MWD's negotiations with IID. Jan Matusak also participated in the negotiations.

Negotiating Positions of the Primary Parties

The four main parties to the transfer—PVID, MWD, the landowners, and the farmer-lessees—each had different interests and faced different risks. In order to annually divert and apply water to beneficial use, and thereby preserve its first-priority water right, PVID needs a well-maintained system of canals, laterals, and turnouts. These are outdoor structures subject to searing heat, erosion from wind, water flow, occasional monsoons, and burrowing by gophers and ants. In some cases, the infrastructure dates back to the nineteenth century. If it falls into disrepair, PVID runs the risk of not being able to divert sufficient water to meet its farmers' needs and possibly over time losing some of its privileged water rights. PVID collected water tolls of $43 per irrigated acre to cover the cost of maintaining the infrastructure. It insisted that in any program involving fallowing, it still must be paid its normal water toll even if its total water deliveries decline.

MWD faced the risk inherent in making payments in advance of delivery of the water. Potential problems were posed by the intervening priorities of IID and CVWD, either of which might claim PVID's saved water. Moreover, MWD's Colorado River Aqueduct has finite capacity. The aqueduct's capacity is 1.3 million acre-feet (maf), and annual flows are near this capacity, as shown in table 6.1. Conveyance of water saved from PVID fallowing would have to be delayed through storage until capacity in the aqueduct was available.

TABLE 6.1

Metropolitan Water District of Southern California's Diversions from the Colorado River, 1985–1994

Year	Diversion (acre-feet)
1985	1,269,526
1986	1,299,650
1987	1,278,836
1988	1,200,181
1989	1,201,390
1990	1,214,971
1991	1,252,352
1992	1,193,830
1993	1,204,003
1994	1,300,203

Source: Metropolitan Water District of Southern California, interview with Fadi Kamand.

Yet another delivery risk concerned long-term storage of the water in one of the Bureau of Reclamation's reservoirs along the Colorado River. Not only is the river fully utilized in terms of diversion claims, but the Bureau of Reclamation must also manage water flows and reservoir levels to control flood risk, meet hydropower commitments, protect environmental values, and meet the delivery and quality terms of U.S. treaties with Mexico. The Bureau's flexibility and willingness to serve as a water bank was by no means ensured.

The interests of landowners and farmer-lessees were aligned for the most part. Both were interested in maintaining the quality of the soil, which could be put at risk in three ways from fallowing. First, without regular applications of water, soil can be degraded when dissolved salts are deposited in the root zone. This occurs through a capillary effect as soil moisture rises to the surface and evaporates. Second, wind erosion can cause loss of topsoil; the valley has 1,300 acres of soil classified as highly erodible. Third, if a fallowed field is untended, undesirable weeds can spring up in the center or on boundaries and eventually deposit seeds over neighboring fields. Managing or eradicating the weeds would then become a multiyear task.

Farmer-lessees also faced the possibility that landowners would require fallowing of otherwise arable land, with the owners keeping the entire water payment. In that case, lessees would have less land with which to earn income. Owners, more so than lessees because of owners' long-term financial stake, also risked creating a perception that by selling water for profit, their full allotment of water was not necessary to their farming activities. This could increase urban or even national sentiment to remove PVID from its catbird seat among lower-basin holders of rights to Colorado River water.

Regarding other potential parties to a PVID–MWD transaction, both IID and CVWD were interested in ensuring that their water use would not be negatively affected by any changes in lower-basin flow regimes. Finally, as described earlier, the Bureau of Reclamation, as manager of the river, wanted maximum flexibility with flow regimes and storage capacity to meet its many obligations.

To summarize, from a contracting perspective, landowners and farmer-lessees would have to sacrifice control over their fields, if only temporarily. The problem for them is that there is essentially no next-highest-valued use for fallowed or unfarmable land in the Palo Verde Valley. A further political asset PVID would have to sacrifice to undertake this transaction would be uninterrupted exercise of its water right. Almost certainly, a major beneficiary of an abridgment to PVID's unquantified first priority would be MWD, the very party with which PVID was contracting. MWD also would benefit if PVID's system of canals degraded to the point that the irrigation district consumed less water. For MWD, the contracting challenge was to craft adequate assurances that water paid for would be saved, stored, and ultimately delivered. Therefore, the district had to design effective oversight of fallowing, measurement of conserved water, storage, and retrieval and delivery when needed. MWD, as a distant urban agency, would face both logistic and informational problems in any oversight efforts.

Regarding initial bargaining positions on price, MWD representatives believed that their district had an upper-limit cost for water of $150/af, the cost at that time for reclaimed water (for application on park grounds or golf courses). Having estimated overhead costs of a fallowing program at $15/af, it determined a highest-offer price of $135/af. PVID farmers believed that their annual break-even price, based on their fixed costs (lease payments, taxes, salaries, water tolls), was $600 per acre. At 4.6 af per acre, this amounted to $130/af. Thus, the bargainers' "reservation prices" created a range of $130–$135/af within which an agreement on price could be reached.

The second time around, negotiations were efficient and productive. In roughly half a year, PVID and MWD teams had reached agreement on all issues and secured approval from their boards of directors. PVID representatives believe that negotiations went smoothly in part because the two sides knew each other and each other's interests as a result of the previous round of unsuccessful negotiations and in part because MWD's Bob Schempp brought an engineer's problem-solving style to the negotiations.[2] The program agreement was signed on May 29, 1992, by representatives of MWD, PVID, CVWD, IID, and the Bureau of Reclamation. Three town meetings were held to explain the agreement to landowners and farmer-lessees, who were then invited to sign standard fallow/compensation agreements with MWD during a June–July sign-up period. The fallow/compensation agreements had been negotiated by PVID and MWD teams concurrently with the program agreement.

The Two Agreements: MWD–PVID and MWD–Landowners and Lessees

The first contract is formally called the "Agreement for the Implementation of a Test Land Fallowing Program and Use of Saved Water" but is referred to as the Program Agreement. The second agreement is the "Agreement for Land Fallowing in the Palo Verde Irrigation District," or the Fallowing Agreement. Each landowner and lessee (if one existed) who chose to participate in the program had to sign an individual Fallowing Agreement with MWD.

The Program Agreement contains twenty-five articles and is twenty-three pages long. The first five pages are recitals that describe the water-rights relationships between the parties and why MWD seeks an agreement. Then the program is tersely described (Program Agreement 1992, Article I):

> The Program consists of a two-year test land fallowing program whereby a specific amount of Saved Water is to be stored by the United States in the Lower Colorado River Basin ("Lower Basin") for use by MWD.... [L]andowners and lessees, in exchange for financial considerations by MWD, are to fallow a portion of the PVID Valley Lands for a two-year period....

In essence, the Program Agreement secures the consent of PVID, IID, and CVWD not to consume their maximum 3.85-maf annual allotment (less Yuma usage) in the years when Palo Verde Valley farmers fallow their fields. By not consuming their full entitlement, the agricultural agencies leave water in the river for a junior appropriator to consume. In this case, the junior appropriator is MWD, with either its fourth and fifth priorities or its surplus-water contract with the Bureau of Reclamation.

The Fallowing Agreement contains sixteen articles and four exhibits, totaling twenty-four pages. In the agreement, landowners and lessees precisely identify their total (or "base") acreage and the fields that will be fallowed and supply a Land Management Plan that identifies how the fallowing will be carried out. On approval of the Land Management Plans, MWD is committed to make five regular payments to lessees of $248 per acre during the fallowing period.[3] This is equivalent to $620 per acre per year or $135 per acre-foot, which falls within the price range described earlier.

Who Actually Owns the Saved Water?

Although the Program Agreement obligates the agricultural agencies not to use their entire 3.85-maf allocation, it does not upset the priority of rights or ownership of rights. For this reason, like the IID–MWD deal, it is not an agreement to transfer water *rights* but is instead a water-*use* agreement. Indeed, the authors of the agreements made sure that the word *transfer* was not used anywhere.

The one element of the transfer that deviates from established appropriative doctrine is that MWD need not put the water to consumptive use in the year it is made available. Instead, the Program Agreement commits the Bureau of Reclamation to store the "saved water" in a lower Colorado River reservoir (Lake Mead) until MWD requests it (Program Agreement 1992). Section 5.1 reads in part:

> Use of the Saved Water by MWD in a calendar year other than the respective calendar year in which the water is saved, shall not reduce the water available to any entity below that which would have so been available absent this Program Agreement.

In other words, the accounting for water use should be made during the fallowing year, and MWD's use happens to be storage. This possibly deviates from "the law of the river" because if MWD envisioned taking ownership of the saved water while at the same time taking its full allocation under its fourth and fifth priorities, it would in total be exceeding its annual allotment. Alternatively, the water might be stored in the form of surplus water to which MWD had rights.[4] But MWD had only a *delivery* agreement with respect to surplus water, not a storage agreement.

In short, there is at best a weak legal basis on which MWD could claim the water not used by PVID farmers in a year other than when it was saved. Because California, along with Arizona, is at the bottom of the chain of users of Colorado River water, the only remaining claimants to that water would be either the river system itself or Mexico. Mexico's rights are spelled out in a treaty, so it does not have a claim on what the United States does with the rest of the water as long as the treaty-specified amount of 1.5 maf crosses the border each year. But the Bureau of Reclamation could use the extra water as emergency backup, to provide additional flows for environmental mitigation or improvement, to generate additional hydropower, or to lower the cost of water delivery by raising slightly the elevation of the river. From a property-rights perspective, the saved water is more the Bureau's water, with a limited delivery option held by MWD, than it is MWD's water, held temporarily by the Bureau.

Key Provisions: How the Parties Handled Transaction Risks

All parties faced short- and long-term risks associated with the plan to transfer water from PVID to MWD. The contracting process reduced the magnitude of some of the risks but not others. PVID sought and received protection for its water tolls: farmers would be required to make full payments on their fallowed lands,[5] and if they failed to do so, MWD would pay PVID on the farmers' behalf and deduct the payments from their compensation.[6] PVID got a further direct infusion from MWD of $250,000 per year to cover marginal costs of managing the program.[7] PVID also received protection of its productive base of soil by means of a requirement for a Land Management Plan. Section 1 of Exhibit D of the Fallowing Agreement (1992) reads in part:

> The aim of the Land Management Plan is to conserve land and water resources and to eliminate or minimize any adverse impacts due to land fallowing on adjacent farms, the community, or PVID.

In terms of protecting its water right, PVID asked for and received two elements: omission of the word *transfer,* as described earlier, and clauses in both contracts stipulating that the agreements "shall not be used or construed as precedent or for argument" in any future negotiations or proceedings.[8]

MWD got commitments from all intervening priority holders (except the Yuma Project, whose 25,000 acres were deemed too small to influence the program) not to use the water saved by the fallowing. Farmers who fallowed fields agreed to undergo extensive monitoring by MWD. Finally, MWD got a commitment from the Bureau of Reclamation to store the water until MWD asks for it or until December 31, 1999.

Even though the agreements identified 4.6 af of water per fallowed acre to be saved from fallowing and storage for use by MWD, MWD's rights to the water had several provisos, the most important of which concerned storage. The Bureau of Reclamation placed two conditions on its agreement to hold water. First, if MWD did not use the water by December 31, 1999, it would lose its rights to it. This created a six-year opportunity to use PVID resources. Second, if flood conditions arose on the lower Colorado River and the Bureau needed additional storage space in its lower-basin reservoirs, MWD's water would be first in line to be spilled.[9]

Next, although the Program Agreement claims that 5.1 af per acre or more was to be saved by the program, MWD was to receive only 4.6 af per acre. Saved water is the water that would have been consumptively used but for the fallowing. In an analysis of the first year of the fallowing program, saved water was measured at 5.1 af per fallowed acre. The 4.6 acre-feet per fallowed acre allocated to MWD was a 10 percent reduction from the measured savings. The 10 percent deduction is taken

> to conservatively provide an assured quantity of Saved Water to MWD, potentially cover evaporating losses in Colorado River system storage and/or develop benefits to Colorado River System storage and/or to all the parties signatory to the Seven Party Agreement . . . , and to facilitate administration of the Program. . . . (Program Agreement 1992, Section 2.2)

Assuming that the figure of 5.1 af per fallowed acre was accurate, MWD representatives felt compelled to relinquish some of the district's water in part to "develop benefits" for the river management system and for other California rights holders, including the Yuma Project. In essence, all of the intervening rights holders were taking small bites out of the apple they were passing from PVID to MWD.

Other risks to MWD's water involved the effectiveness with which the program would be implemented. For example, if immediately after the two-year fallowing period, PVID farmers applied more water to their fields than they historically would have in the same August–December period, that amount could be deducted from MWD's portion.[10] A further possibility of reduction in saved water arose when water was applied to the fallowed fields, whether inadvertently, as a dust-control measure, or in breach of agreement. MWD was to be apprised of such applications and could reduce payments to lessees at the rate of $135/af. Although the amounts of water applied to fallowed fields were to be reported to MWD, the Program Agreement does not specify that this amount would then be deducted from MWD's total quantity of saved water. The deduction would probably occur, however, because that would be the only way to avoid double-counting of applied and saved water. In sum, there were restrictions on MWD's right to the saved water even beyond whether room could be found in the Colorado River Aqueduct to convey it to the coast.

With respect to landowners and their lessees, in order to participate, both parties were required to sign the same Fallowing Agreement with MWD. This reduced the possibility that poor coordination or deception might block either party from executing duties or receiving benefits from the program. Payments were to be made to lessees, consistent with the underlying legal arrangement that MWD was paying not for water rights but for fallowing behavior. How the payments were to be shared among landowners and lessees was left to their mutual agreement.[11]

The other agricultural agencies joined PVID, MWD, and the Bureau of Reclamation on a Measurement Committee.[12] The committee, which operated by consensus, was charged with the tasks of (1) reviewing the status of the fallowed acreage, (2) calculating the amount of saved water available to MWD, and (3) estimating the amount of water actually saved by the agreement. The consensual nature of the decision making, combined with the key task of measuring saved water, gave the agricultural agencies leverage to protect their interests in maintaining a beneficial flow regime.

MWD agreed to prepare a comprehensive report on the program to include descriptions of the acreage fallowed, amounts of saved water, effects on groundwater and cropping, "and any other information necessary to develop a general conclusion on the success of the Program and any impacts that it may have caused."[13] This is a commitment to an extensive and inclusive analysis of the program's effects.

To summarize, all parties got what they wanted, with MWD getting both the least and the most. MWD got the least favorable security features of all the signatories. It was able to negotiate only weak and temporary ownership rights to the water it was to pay for. Moreover, the agreement left MWD facing a number of other risks: that the Measurement Committee's consensus process might break down, that monitoring and enforcement of farmers' compliance with their land management plans could prove difficult or contentious, and that post-1999 storage of saved water in the lower basin could prove problematic.

On the other hand, MWD got the most symbolically for reaching a voluntary water-transfer agreement with an irrigation district and its farmers, and with none other than PVID, which for more than a century had thrived on anonymity while sitting at the pinnacle of water-use priorities for the lower Colorado River.

Executing the Agreement

At PVID's three town meetings, at least two landowners expressed serious initial reservations about the program. One, San Diego Gas & Electric (SDG&E), owns land in the district because it had planned in the 1970s to build a nuclear power plant west of the valley. SDG&E's primary region of activity is in MWD's service territory. With the plant still unbuilt,

SDG&E's valley land was regularly farmed by a lessee. Now SDG&E faced the prospect of paying twice for the same water. In addition to its normal water bill, SDG&E would also have to help pay PVID farmers to fallow their land. If SDG&E's own lessee chose to participate, it would amount to SDG&E paying its lessee for water SDG&E already owned. Despite this double-billing concern, SDG&E's lessee eventually participated in the program under the same terms as everybody else.

Another concerned party was Desert Security Farms, Inc., a farming company with operations throughout the United States. Desert Security, which is wholly owned by the Church of Jesus Christ of Latter-Day Saints (Mormon Church), had recently experienced difficulties with a water-marketing program in Florida and was suspicious of any such program elsewhere. Eventually, however, Desert Security enrolled in the program. Other concerns were raised by owners of businesses that served the agricultural economy, including Waymen Dekens, owner of Dekens Implement Company in Blythe.

Ultimately, sixty-three fallowing contracts were executed and 20,215 acres were fallowed, 22 percent of PVID's total cropped acreage. Two hundred sixty-one separate fields were chosen for fallowing, for an average of 77 acres per fallowed field. The fallowed area fell only 1,500 acres short of MWD's original vision of saving 100,000 af of water per year but was otherwise consistent with the agreements' guidelines on fallowing. In some cases, farmers with too little land to meet minimum acreage requirements voluntarily joined together to meet the requirements and participate. Figure 6.1 shows that fallowing was relatively well distributed throughout the valley. Fewer than five farmers chose not to participate.

The agreements were executed smoothly. Very few issues arose. Among them were some unexpected difficulties in identifying an owner for a few of the farms because titles resided in complex trusts or other family arrangements. There also was an unusually wet winter just after the program commenced: 9.2 inches of rain fell in the first year, compared with a historical average of 4.2 inches. The extra rain caused weeds to spring up in fallowed fields, and these had to be removed at additional cost to the farmers.

Neither the city nor businesses nor labor interests were apprised of the transfer negotiations. The city manager learned of the transfer informally and alerted the city council. Once the program was operating, PVID's role diminished significantly to its usual activities of providing water, maintaining infrastructure, and performing billing and accounting duties. The main

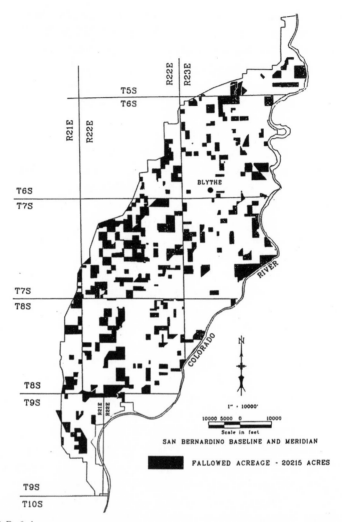

FIGURE 6.1

Location of Fallowed Acreage, PVID–MWD Test Land-Fallowing Program

Source: Great Western Research, Inc., "Palo Verde Test Land Fallowing Program, August 1, 1992–July 31, 1994, Final Report," vol. 1 (Report prepared for the Metropolitan Water District of Southern California, August 1995).

activities of the program were carried out by farmers, MWD, and MWD's local representative.

Three key aspects of program implementation were how the fallowed fields were chosen, how the fallowing was carried out, and how MWD dealt with the challenge of monitoring and enforcing the fallowing agreements. Farmers were given the right to choose which fields would be fallowed, with the following conditions: No more than 25 percent of their base acreage could be fallowed. Base acreage had to have a history of agricultural production since 1987. The cap of 25 percent per farmer was suggested because fallowing of that portion of land in PVID would generally yield MWD's goal of 100,000 af per year of saved water. Further, PVID negotiators believed that capping each farmer's participation at 25 percent would give all farmers an equal opportunity to participate in the program, without domination by owners of a few large farms. Fallowing could take place only on plots of at least eighteen contiguous acres. The eighteen-acre minimum requirement was meant to ease monitoring and enforcement of the fallowing.

Farmers knew their land and selected their least productive lands for fallowing. There is a gradation in quality of soil in the valley, primarily due to various levels of sandiness and salinity. The choice of low-quality soils for fallowing resulted in a lower percentage of loss of production than of acres fallowed. In estimating the loss of production, Butch Hull (1994), public works director of Blythe, stated that in general, a 50 percent fallowing of valley lands would cause a 35 percent loss of production, and that this fallowing program had caused a 15 percent reduction in production for the valley overall. Crop consultant Garn T. Stanworth (1995) similarly estimated that a 25 percent fallowing would lead to a 10–15 percent decline in production. He pointed out that the highest-quality soil will yield an alfalfa harvest of roughly 14,000 pounds per acre, whereas the lowest-quality soil will yield 6,000–7,000 pounds per acre. Therefore, an agreement that allowed individuals with detailed knowledge about the land to make least-cost decisions about which fields to fallow resulted in selection of the fields with the lowest economic impact.

In response to a survey, farmers estimated that 64 percent of the fallowed fields would have been planted with alfalfa. Alfalfa was a logical crop to forgo because it is more likely than vegetables to thrive in low-quality sandy or saline soils. In terms of the fallowing agreement, two complementary qualities of alfalfa are that it is extremely water-intensive and that it requires almost no seasonal labor input. Thus, when an alfalfa field is fallowed instead of a row-crop field, more water is saved and fewer farm-labor jobs are lost.

During the fallowing, 1,250 acres of highly erodible soils required special treatment. This involved spreading grain seed over the fields and relying on rainfall for germination. The plants were allowed to develop mature root systems and then were chopped, mowed, or swathed. The remaining root systems and plant residue protected the soils from wind erosion. The U.S. Department of Agriculture's Soil Conservation Service served as advisor and overseer of protection measures for the highly erodible soil.

Responsibility for implementing, monitoring, and enforcing the fallowing was divided between PVID and MWD, with the bulk falling on MWD. PVID had installed a computerized water-delivery system that could report whether fallowed fields were inappropriately receiving water. MWD was responsible for on-site monitoring and enforcement. To carry out this task, it hired a longtime resident and local farmer, Bob Micalizio. With forty-five years of farming experience, Micalizio was well known in the valley. Nicknamed "the water cop" during the fallowing period, he regularly visited all the fallowing sites and soon knew all the farmers by name. Although he experienced few problems in enforcing the agreement, he was aware of reticence on the part of some farmers to maintain irrigation ditches that were not being used as a result of the fallowing. Micalizio also encountered difficulties with some absentee farmers. There were a few cases in which an absentee farmer did not in a timely manner take out weeds that had sprung up on a fallowed field. Micalizio disked the weeds himself in those instances and then sent a bill to the farmer. He never encountered hostility toward himself as an MWD representative or toward MWD.

By the conclusion of the fallowing program, MWD's overhead costs were $10 per af saved, well below the $15/af estimate. MWD had managed to delegate authority for monitoring and enforcing its agreements to a credible and effective local representative. Farmers were satisfied with the program, and many said that at an appropriate price, they would be willing to participate again. MWD also was satisfied with the fallowing portion of the program.

Effects of the Agreement

The two-year test land-fallowing program concluded on July 31, 1994. PVID experienced few negative effects. It had a new computerized water-tracking system. Its canals and ditches were intact, and it remained in solid financial condition. A Bureau of Reclamation initiative to quantify all water rights in the lower Colorado River basin, including PVID's, arose but soon became bogged down, so there appears to have been little political fallout from the decision to transfer water. Instead, the state's interest in promoting water transfers has remained focused on the San Francisco Bay/Sacramento–San Joaquin Delta system. The other agricultural agencies, IID and CVWD, also appeared to be satisfied with the program.

With respect to farmers, the infusion of cash from MWD helped carry a number of them through difficult financial times associated with low crop prices and pest infestation. According to a 1995 MWD-sponsored survey of farmers carried out by Great Western Research, Inc., 43 percent of MWD's first-year payments and 35 percent of its second-year payments went to loan repayment at local and out-of-town banks. Additional funds were spent on farm improvement, farm operations, and other miscellaneous items.

In terms of soil conditions, after two years of fallowing, the capillary effect associated with drying soil had pulled additional salt up into the root zone, but the salt was washed away with one preirrigation watering. Farmers initially reported to crop consultant Garn Stanworth in 1995 that they saw improved yields on previously fallowed lands. However, a February 1995 survey of farmers suggested that the two years of fallowing generally had no effect on soil productivity. The risk of erosion on the valley's 1,250 acres of highly erodible soils was largely mitigated by the special fallowing methods described earlier. Finally, the risk of deposition on neighboring fields of undesirable weeds that grew in untended fallowed fields was largely mitigated by careful attention to complete fallowing.

In terms of labor effects associated with the fallowing program, local representatives of the California Employment Development Department (EDD) noted that during the fallowing period, many workers did file for unemployment benefits, saying that there was not much work out there. According to one EDD employee, the fallowing program had compounded the adverse effects on the labor force brought about by the whitefly infestation. Local labor leaders recounted the stories of two tractor drivers who had to take lower-paying jobs because tractor work was less available during the fallowing period.

The survey of farmers carried out by Great Western Research in 1995 revealed minimal, though concentrated, labor effects. During the first year of the program, farmers reported a loss of jobs representing 14.5 full-time-equivalent equipment operators, 10 of which were associated with a single "custom operator" with 1 percent of the fallowed acreage. Also in the first year, 11.5 full-time-equivalent irrigator jobs were lost. Given the EDD estimate of 800 full-time agricultural workers in the valley, this represents a 3 percent reduction in permanent farm employment as a result of fallowing in the first year.

In the second year of the program, 1 full-time-equivalent equipment operator job was reported lost and no full-time-equivalent irrigator jobs were reported lost. The farmers reported no job reductions for seasonal equipment operators or irrigators or for full-time or seasonal laborers in either year as a result of the fallowing program.

In terms of subsequent labor effects, Stanworth noted that during the fallowing period, farmers discovered additional ways to run their farms economically, including more efficient use of labor. However, though this would tend to reduce labor demand over time, there was also an increase in vegetable production in the valley, which increased labor demand.

The city of Blythe experienced few effects from the transfer. There did not seem to be a change in demand for city services as a result of increased unemployment. The city manager, Les Nelson, believed this was because the fallowing of alfalfa fields had minimal labor effects. The town may also have been insulated from adverse effects on labor by the infusion of new jobs and families from construction of the two state prisons.

Many businesses in Blythe that are linked to the agricultural industry were at first opposed to the fallowing. One outspoken opponent of the program was Waymen Dekens, owner of Dekens Implement Company. After an initial reduction in sales, however, Dekens saw increased demand for farm implements and repairs, and as of March 1995, he was expanding his store. Among businesses that suffered during the fallowing period were custom alfalfa sellers and a boarding house for farmworkers. The house has a thirty-person occupancy and usually has a waiting list, but during the fallowing period, there were fewer occupants and shorter stays. The owner of the boarding house, Jose Fernandez, estimated that occupancy declined by 30 percent during the fallowing period.

In terms of effects on California's southern coast, MWD was able to demonstrate that it could enter into a fallowing agreement with an irrigation district and its farmers. However, it could not demonstrate that water would flow its way as a result. The Achilles' heel of the agreement—MWD's weak storage agreement with the Bureau of Reclamation—eventually led to the water being returned to the river. By January 1997, the Colorado River basin had already received heavy precipitation and had a deep snowpack, and the National Weather Service was forecasting more. To prepare for potential flood conditions on the lower Colorado, the Bureau started releasing water from Lake Mead and other reservoirs, and MWD's water was first in line. Between January and March, MWD's $25 million investment in saved water splashed down Hoover Dam's spillway and flowed on to Mexico.

Unique Elements of the Agreement

Before distilling lessons from the Palo Verde experience, a discussion of what is unique and probably not replicable in the agreement is appropriate. Unlike the situation in other parts of the state, only a handful of parties had water rights that could have been affected by the fallowing agreement: the five signatories, the Yuma Project, and Mexico. The hydrology of the Yuma Project (direct access to the river) and its relatively small demand protected its interests, and Mexico's interests were secured by its treaties with the United States. All parties had the financial resources necessary to devote years of attention to negotiation and implementation of the agreements.

Physical impediments to additional deliveries reduced the threat posed by MWD's combination of high demand and deep pockets. The Colorado River Aqueduct carries no more than 1.3 maf/year, and MWD already had a firm right to half that amount and a contingent right to the balance. Without additional expensive conveyance structures, MWD could not make off with the agricultural agencies' water. Farmers in the Palo Verde Valley were therefore less subject to the "Owens Valley syndrome," that is, the belief among farmers and residents of rural towns that even a small agreement with a large urban buyer could lead to complete loss of a region's water rights. Further, the valley has a low level of absentee farm ownership, and even when a plot is leased, the owner is usually a retired farmer who still lives in the valley. This helps align the concerns of all the different farming interests (farmer-lessees, farmer-landowners, and landowners) and makes valleywide agreements more feasible.

The valley's location in the midst of the Sonoran Desert reduced potential effects of the fallowing on any communities besides Blythe and the few other smaller towns in the valley. The city of Blythe had already taken action to attract nonagricultural jobs to the area and was less dependent on agricultural income than at any other time in its history. Because labor was organized only at citrus farms, where fallowing was out of the question, there was little potential for organized-labor opposition to a fallowing plan. Indeed, the language barrier between the predominantly Spanish-speaking workforce and English-speaking landowners and farmer-lessees, exacerbated by the lack of a Spanish-language newspaper in Blythe, inhibited the labor sector's awareness of the transfer negotiations.[13]

By 1991, the valley was suffering from low commodity prices and a whitefly infestation. These commercial and biological challenges combined to provide a strong incentive for farmers to consider ways to generate revenues from their water resources. Given the valleywide conditions, it was fortuitous that the contract left fallowing decisions to individual farmers, each with a 25 percent cap on total acreage. This allowed them to select the lowest-quality soils for fallowing. Alfalfa production declined, with the combined benefits of high amounts of saved water and minimal adverse effects on labor. The effect on the labor force of a 22 percent fallowing of the valley may have been as low as a 3 percent reduction in full-time-equivalent permanent jobs. In executing the agreement, MWD took the innovative step of identifying a local individual, Bob Micalizio, himself a former valley farmer, to represent its interests. Without a trustworthy, non-confrontational, low-cost means of monitoring and enforcement (i.e., the water cop), the agreement would have been much more costly to implement and perhaps more contentious.

Finally, just as in the IID–MWD transfer, the appropriative-rights priorities and the pattern of demand along the Colorado River created an opportunity to transfer water without transferring water rights even temporarily. The senior appropriator simply let the water "flow down" to the junior appropriator in exchange for a financial consideration. The PVID–MWD transfer was never formally reviewed by the State Water Resources Control Board, which is typically empowered to review water transfers in the state.

Lessons and Provisos

Unlike the situation in the IID–MWD transfer, in which CVWD took independent legal action to block the transfer and then to exact concessions, this time MWD convened all the major intervening priority holders (except for the relatively small Yuma District) and made an advance agreement. MWD received commitments from the intervening parties not to use the saved water, but it essentially had to donate 10 percent of what was saved in order to get the commitments. MWD also agreed that the parties would form a committee to determine the quantity of saved water, with each party holding veto power over measurement decisions with which it disagreed.

A risk to PVID, as well as to landowners, was that interruption of more than a century of regular annual water use could lead to erosion of PVID's enviable water-rights position. The agreement could have been characterized as an admission by PVID that it did not need all the water to which it was entitled. This in turn might have increased pressure to reform the

Seven-Party Agreement governing California users of the Colorado River or to reform federal rights to Colorado River waters.

One seemingly obvious way to deal with this hazard was for PVID simply not to sacrifice its historical pattern of water use, even for one year. But this strategy had its own risks. PVID might have been seen as insensitive to growing urban needs or to urban suffering during a drought. Such perceptions also could have led to a revisiting of Colorado River allocations, in which PVID would have been the only party with nothing to gain.

PVID ultimately chose a short-term test-transfer format and built in two contractual defenses against future calls for reallocation. The first is that the word *transfer* appears nowhere in the contracts, thus reducing the perception that PVID had extra water to sell. Second, the fallowing agreement contains "no-precedent" language that attempts to minimize the legal significance of the transfer. The no-precedent language attempts to characterize the fallowing program as a nonevent with respect to reallocation. Because no significant move has been made to downgrade PVID's rights in the years since the agreement, in the short run the transfer has not eroded the stability of PVID's rights. Potential long-term effects, if any, are more difficult to gauge.

Cumulative effects, such as those pertaining to farm labor, agriculture-related commerce, and soil quality, were measured over a space of only two years during the fallowing. It would be risky to argue without qualification that adverse effects of any proposed longer fallowing program would be equivalent to the minimal adverse effects of the PVID–MWD transfer. For example, some farmers were resistant to maintaining idled ditches to protect them from burrowing or other damage even though they knew the program would last only two years. This resistance might increase in a longer fallowing program, leading over time to expensive damage to an irrigation district's infrastructure.

In conclusion, many of the contracting risks associated with the fallowing program were met with solid responses that reduced or eliminated potential problems. Some mechanisms were quite effective in mitigating potential hazards, such as designation of a local farmer to look after MWD's interests. Others were ineffective, such as the Bureau of Reclamation's temporary, conditional storage commitment to MWD. The many unique attributes of the transaction notwithstanding, this transfer suggests that fallowing arrangements are feasible in California but (just as with the IID–MWD transfer) are likely to involve complex, long-term contracting.

Notes

1. This figure, 4.6 af of water saved per acre, differs from 3.6 af of water utilized per cropped acre (mentioned on page 66) because the same acreage may be cropped more than once per year in PVID, raising overall use per acre.
2. Two PVID representatives said that they found working with Schempp, an engineer, more productive than working with Jan Matusak, a detail-oriented attorney, because the presence of an attorney made them uncomfortable. Matusak, it turned out, is also a professional engineer, but obviously one who is careful about details.
3. Fallowing Agreement 1992, Exhibit D-1, Section 4.a.
4. It is possible that MWD could claim ownership of the saved water under its Surplus Water Contract with the Bureau of Reclamation. Section 14.1 of the Program Agreement asserts that provisions of MWD/Bureau 1987 Contract No. 7-07-30-W0171 for Delivery of Surplus Flows have "the same force and effect" as if they were spelled out in the Program Agreement.
5. Fallowing Agreement 1992, Section 3.f.
6. Ibid., Section 5.b.
7. Program Agreement 1992, Section 3.2.
8. Ibid., Article 12, and Fallowing Agreement 1992, Section 14.
9. Program Agreement 1992, Section 2.1.
10. Ibid., Section 2.3.
11. Fallowing Agreement 1992, Section 3.i.
12. Program Agreement 1992, Section 4.3.
13. Ibid., Section 4.5.
14. The *Palo Verde Times* had a Spanish-language page on a weekly basis.

CHAPTER 7

Water Ranching: The Devil's Den Water District–Castaic Lake Water Agency Combined Land Purchase and Water Transfer

Item, September 30, 1955, 5:45 P.M. Actor James Dean, driving "Little Bastard," his Porsche 550 Spyder, dies in a car crash near the isolated town of Cholame, California. Cholame lies just west of the Devil's Den.

The Devil's Den is a small shelf among the Pyramid Hills about two miles southwest of Devil's Den Ranch and the Devil's Den Water District. This desolate, undeveloped region, which rides the border of Kern County and Kings County on the western side of the San Joaquin Valley, has an elevation climbing west from 400 to 500 feet at a rate of 10 to 15 feet per mile. The roughly 10,000-acre region once was full of deer and elk and had a year-round water source in Cottonwood Creek, which flowed from west to east through the Dagany Gap. The endangered San Joaquin kit fox still roams the region.

Before groundwater-based farming became established in the 1930s, groundwater levels there were very high. In a recent survey of existing wells in the district, the well marked number 1, presumably the first well drilled, was found to descend only forty-three feet. In the 1980s, well depth in the district ranged from 900 to 1,700 feet. Groundwater generally flows from west to east through the Dagany Gap, draining a 500-

square-mile region of rolling hills and mountains. Some additional groundwater moves south through the Avenal Gap. The groundwater is extremely
saline. Although the aquifer under the Devil's Den Water District is hydrologically separate from other aquifers, it shares the poor water quality that
renders groundwater for another sixteen miles east of the site unsuitable
for drinking or irrigation.

Irrigated farming began at Devil's Den in the 1930s when two Irish
brothers drilled the first water wells and planted a barley crop. Ownership
changed hands occasionally over the years until Producers Cotton Oil
Company bought Devil's Den Ranch in 1957. (One of the interim owners was Cecil White, who, as a U.S. senator from California, cosponsored
the 1928 Boulder Canyon Project Act with Congressman Phil Swing of
Imperial County. The first element of that act was the building of the All-
American Canal, followed by construction of Hoover Dam.) Producers,
through its landholding subsidiary Southlake Farms, also farmed acreage in
Fresno and Tulare Counties and by 1959 had 90,000 acres under cultiva-
tion, roughly 9,400 acres of which were located at Devil's Den Ranch.

As shown in figure 7.1, the irrigation district is shaped like a sideways
T. It has a main north–south corridor and then pushes west, back into the
hills, through the relative lowlands of the Dagany Gap.

Irrigated farming at Devil's Den Ranch was accomplished using well
water exclusively (no imported water) until 1970, and some dryland farming took place in the western portion of the ranch. One main ditch moved
water north and south from the midpoint of the ranch, which also formed
a high point in elevation for the eastern side of the property. Some small
pumps lifted water west onto fields higher in elevation.

The Devil's Den Water District (DDWD) was not formed until 1965.
At that time, the State Water Project (SWP) was looking for contractors for
water to be delivered along the California Aqueduct. Although a great deal
of DDWD's land is in Kern County, Producers secured its own separate
contract with the SWP rather than subcontract through the Kern County
Water Agency, as nearly all other county water agencies did. Producers
formed an 8,675-acre district to serve its own property and then contracted for 12,700 acre-feet (af) per year of agricultural water. To create a
board of directors for its water district, it transferred title to some plots of
land on the ranch to some of its employees and then named them to the
board!

Delivery to DDWD was expected to be relatively simple because the
California Aqueduct's coastal branch, slated to serve Santa Barbara, would
pass right through the district. Construction of the branch was started and

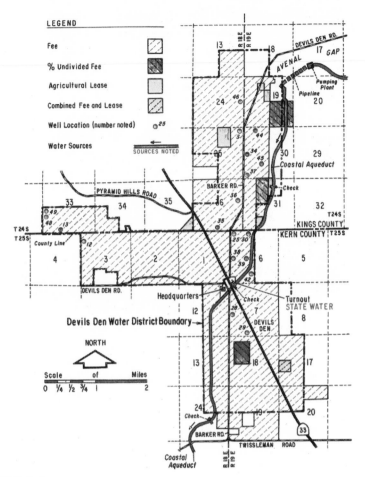

FIGURE 7.1
Devil's Den Ranch

then stopped, causing it to become known as the "coastal stub." Luckily for Devil's Den, the stub carried water right to its doorstep. SWP deliveries began in 1970, and in some years, DDWD was able to take not only its 12,700-af allotment but also 1,000 to 1,500 af of surplus water at no charge.

At the time deliveries began, Producers decided not to invest in expensive new infrastructure for water delivery and application, something nearly every other SWP contractor was doing, including its next-door neighbor, the Berrenda Mesa Water District. Instead, it would simply lift the water into the same ditches that had been carrying well water since the

1930s. The process would be inefficient, but it would not place heavy long-term debt on the farm for a new delivery system.

Cotton growing faced a number of challenges at Devil's Den Ranch. Although the soil quality was excellent, groundwater quality was not, and repeated applications of saline water increased the salt content of the soil. The sloping hillsides made furrow irrigation more difficult because water often would not run slowly down furrows, sinking into the soil, but would rush to the far end of the field, creating an erosion problem. Erosion also was evident in the main water-supply ditches, which were unlined. The erosion problem could be handled, but only with capital-intensive use of drip- or sprinkler-irrigation systems. Moreover, the region's 400-foot increase in elevation from the valley floor meant that growing seasons for cotton were slightly shorter than at the ideal (zero) elevation, again not favoring competitive cotton production. In spite of these hindrances, cotton was successfully farmed there year after year.

During the height of the ranch's productivity, in the 1960s and 1970s, Producers assembled eight World War II surplus army barracks to house employees and families, as well as some additional small buildings for unmarried and seasonal workers. The farming was labor-intensive: the combination of hand-moved sprinkler and furrow irrigation and the picking of cotton by hand required a staff of roughly fifty full-time employees as well as additional labor at harvest time. There were no towns in or near the district; the closest ones were Avenal, with fewer than 3,000 residents, about eighteen miles north along Highway 33; Corcoran, nearly an hour's drive northeast; and almost nothing to the west until Cholame and then Paso Robles, on the other side of the Diablo Mountains. A proposed Devil's Den town site, which had been subdivided with lots sold off in the 1920s or 1930s, had been no more than a mineral-rights speculation. Under the guise of city planning, lots over what was thought to be an oil field had been parceled and sold but with no intention of building a town.[1]

Cotton was always the main crop at Devil's Den Ranch. Producers was a large, vertically integrated company that owned cotton-growing farmland, cotton gins (large machines that separate cotton seeds from cotton lint), and cottonseed-processing equipment for producing such products as vegetable oil, livestock feed, and cellulose. Producers packaged and shipped the products to wholesalers and end users. In addition to maintaining its own farmland, Producers provided crop insurance to private cotton farmers and purchased their crops annually. By the mid-1980s, Producers was ginning as much as 240 million pounds of raw fiber, or 500,000 bales of cotton, an amount equivalent to the average output of four of the country's fourteen traditional cotton-growing states. Despite these numbers, employment demand at Devil's Den Ranch had declined. By the mid-1980s, roughly thirty people were still full-time employees, down from about fifty, with cropped acreage holding at approximately 5,000 acres per year. Profit margins were dwindling.

Producers was soon purchased by a conglomerate called Bangor Punta Corporation. This firm had originally owned a sugar operation in Cuba, and after the operation was nationalized, it used U.S. tax credits to purchase such firms as Smith & Wesson, Piper Aircraft Inc., and Producers. In the early 1980s, Bangor Punta sold Producers to Lear Siegler, Inc., which in the mid-1980s sold Producers to Dunavant Enterprises, Inc., one of the world's largest cotton merchants. Dunavant, based in Memphis, Tennessee, was not in the cotton-*farming* business. A Producers brochure from the Dunavant period never mentions farm ownership as an activity of the company, instead stressing "financing a farmer's production, providing competitive prices and guaranteed quality for cotton fiber, and assuring timely delivery of prime vegetable oils or high grade cotton to far away customers" (Producers Cotton Oil Company n.d.).

At least three factors suggest a reason why Producers became interested in selling the Devil's Den Ranch property. First, the economics of farming differ from the economics of ginning, processing, packaging, and shipping cotton products. Producers found that in the 1980s, its investment in land was not yielding the same returns as were other corporate activities, and it decided to sell all its landholdings. Until that time, Devil's Den Ranch had been a "cash cow" for Producers, requiring little capital investment and yielding high returns. Now, though not a money loser, it was not as profitable as before. Second, and similarly, with the rising costs of power, it was becoming more expensive to lift water first out of the California Aqueduct and into DDWD and then up again onto fields, and profit margins were shrinking. The earlier decision not to build a new delivery infrastructure when the SWP water arrived meant that the existing conveyance systems would need an expensive upgrade if a long-term commitment to farming were to be made. Finally, by the time Dunavant purchased Producers, it clearly wanted all its remaining landholdings sold, since it saw itself as a cotton merchant rather than a cotton grower. Dunavant was a motivated seller of the property.

In preparation for negotiations, Producers hired Summers Engineering, Inc., to examine the property and determine what could be of value. (Joseph Summers, owner of the firm and a highly respected water engineer in California, would soon play a role in the IID–MWD conservation and transfer agreement, serving as chairman of its Program Coordinating Committee.) Summers concluded that the water contract might be more valuable than the land. Thus, although the goal of Producers was to sell the land, company executives knew from the beginning that the water contract could be what brought the property the majority of its value.

Castaic Lake Water Agency: Fueled by and Fueling Southern California's Growth

Just northwest of the San Fernando Valley in southern California lies the smaller Santa Clarita Valley. Once a site of citrus farms and poultry ranches, it has slowly been transformed into the northern residential-commercial boundary of Greater Los Angeles. Most of the valley lies in Los Angeles County; some of the western portion lies in Ventura County.

In 1962, the Castaic Lake Water Agency (CLWA) was formed with the principal purpose of entering into a contract for State Water Project deliveries (an origin similar to that of DDWD). CLWA acquired an entitlement of 41,200 af/year, which it treats and delivers to the region's four water agencies: Los Angeles County Waterworks District No. 36, the Newhall County Water District, the Santa Clarita Water Company, and the Valencia Water Company. As a water "wholesaler," CLWA plays the same role as does MWD, but on a far smaller scale.

According to Bob Sagehorn, general manager of CLWA, the amount of water specified in the original contract with the SWP was based on what the region could afford to pay, not on a projection of population growth. In the late 1970s, the district's population was expected to double over the next two decades to 164,000 residents, requiring far more water than the SWP entitlement, occasional SWP surplus water, and local supplies could reliably provide. Motivated by their four member agencies, whose concern was a reliable water supply for an expanding region, CLWA began actively seeking water to purchase around 1984 (Sagehorn 1995).

Negotiations: Two Rounds

The first meeting between Producers and CLWA took place at a meeting of the Association of California Water Agencies (ACWA) in 1984. By that time, Producers had sold its other landholdings in Fresno and Tulare Counties, and only Devil's Den Ranch remained. Bob Sagehorn and the rest of the staff had spread the word that CLWA was looking for water to purchase, and he soon began speaking with representatives of Producers about the Devil's Den Ranch property. Agricultural property values were high at the time, and Producers was asking roughly $3,000 per acre, or about $30 million, for the property. This implied a permanent water transfer price in the range of $2,300/af, which was well beyond the range CLWA was interested in paying. At that same ACWA meeting, Producers held informal meetings with both MWD and the San Diego County Water Authority about the possible sale of Devil's Den Ranch. None of the three potential buyers was interested enough at that price to pursue matters. CLWA representatives also spoke with other potential sellers of water, with no results.

Two changes in circumstance brought Producers and CLWA back to the table in 1987. First, prices for agricultural land were falling, so the original asking price of nearly $30 million was no longer realistic. Second, Producers had been acquired by Dunavant. Dunavant ordered the immediate sale of the Devil's Den Ranch property, permitting its negotiators to accept a deeply discounted price. Producers contacted CLWA, and the two sides again sat down. This time, negotiations went on for only six weeks before an agreement was signed. At the time, very little outside attention was paid to what Producers described as an effort to sell its Devil's Den Ranch property.

The Agreements: An Option to Purchase Land and a Side Agreement to Facilitate the Water Transfer

While this book was being researched, CLWA's historical documents were in boxes in storage as a result of condemnation of the agency's headquarters following the 1993 Northridge earthquake. Meanwhile, Producers had changed hands so often through the 1980s and 1990s, ultimately becoming part of the Anderson Clayton Corporation, that neither side could furnish a copy of the contract. (One attorney involved in the negotiations chose not to furnish a copy of the agreement because he thought that public knowledge of its provisions would harm his ability to expand his water-transfer business.) Nevertheless, key elements emerged during interviews with the negotiators and reviews of related documents.

The primary agreement took the form of an option to purchase essentially all of the interest held by Producers in lands served by DDWD (roughly 90 percent of the district) and 700 additional acres adjoining the district. The sale included transfer of existing farm leases on plots not owned by Producers (at the proposed town site) as well as farm equipment and equipment leases. One exception was the cotton gin and its surrounding land, ownership of which was to be retained by Producers. The option itself cost $250,000, which CLWA provided on signing the agreement, and the purchase price was set at $5 million. This amounted to one-sixth the original asking price of nearly $30 million, a deep reduction reflecting both the decline in farmland values and the extent to which Dunavant was motivated to sell the last parcel owned by Producers. This comes to roughly $400/af of permanently transferred water.

In a side agreement, Producers, which completely controlled DDWD and its board of directors, agreed to set up a new joint powers authority, a public agency formed by and between DDWD and CLWA, which CLWA would control. This new agency was called the Devil's Den–Castaic Lake Authority, and its primary purpose was to complete the state-mandated environmental impact report (EIR) describing potential effects of transferring the water away from Devil's Den Ranch.

For Producers, this was no more than an agreement to sell property, and as such it required no additional risky investments. The only risk was that if the purchase option was not exercised, Producers would have to look for another buyer. Presumably, the option's purchase value of $250,000 covered the cost of this risk. For CLWA, the primary risk was that although the agency would be able to purchase the land, it might not subsequently be able to transfer the water.

Executing the Agreement

For the water transfer to be completed, two goals had to be achieved. One was the satisfactory completion of an EIR. The other involved amendment of DDWD's and CLWA's contracts with the SWP to recognize the transfer of DDWD's entitlement to CLWA. Achievement of the second goal depended in part on achievement of the first goal. This work was done solely by CLWA and its counsel and consultants; Producers took no part in these efforts.

The draft EIR discusses the environmental impacts of the transfer of all or a portion of the 12,700-af/year DDWD entitlement to CLWA. The transfer was expected to take place during the mid-1990s, according to demand projections for CLWA's service territory. The draft EIR noted that in 1988, only 3,500 acres were under irrigation out of the 9,400-acre total, and that farming could continue in the area using groundwater on a sustainable basis. The reversion to sustainable groundwater farming was seen to be environmentally beneficial, compared with intensive cropping using imported water, because of the reduction in the amount of pesticides and herbicides introduced, reductions in erosion, an increase in natural habitat, and potential reestablishment of native grassland. There was little discussion of the effects the water transfer would have on the environment in CLWA in terms of the type and scope of development the water would support.

Copies of the draft EIR and its Notice of Completion were sent to agricultural and urban water agencies; county government agencies in Los Angeles, Ventura, Kern, and Kings Counties; federal and state agencies; members of the boards of directors of both DDWD and CLWA; and engineering firms that had worked on the transfer. No environmental or farm-oriented nongovernmental organizations were directly apprised of the transfer, although notice was placed in three newspapers that the forty-five-day comment period had begun on July 8, 1988.

Only three substantive comment letters were received, all of them from public bodies. They came from the California Department of Conservation, the California Department of Food and Agriculture, and the Kern County Department of Planning and Development Services. Although few in number, these comments, along with CLWA's responses in the final EIR, provide a public record of a dialog on the merits of transferring water from rural to urban uses. Essentially, if CLWA could answer the concerns raised by the commenters in its final EIR, it could have the water. If not, the water was likely to stay at Devil's Den Ranch.

A central concern expressed in all three letters was variously described as the "loss of valuable farmland," the loss of "prime farmland" as defined by the U.S. Department of Agriculture, and the loss of lands contracted under the California Land Conservation Act (Williamson Act).[2] The joint powers authority provided a number of answers to this concern (Devil's Den–Castaic Lake Authority 1988):

- The land will remain potentially irrigable in a physical sense, and ready to return to production if a new imported water source is someday found to serve it. This is a water transfer to an urban area, not a transformation of rural land to urban uses.
- Farming in a form similar to farming between the 1930s and the 1960s is still possible at Devil's Den Ranch, especially if salt-resistant barley becomes the main crop.
- Mitigation measures could reduce the possible adverse environmental impacts of fallowing fields. These could include soil and water conservation surveys, reseeding with native grasses, and weed control measures.

The other primary concern, expressed in the letter from the Kern County Department of Planning and Development Services (p. D-2), has to do with "the actual cumulative effect of this project and other [water-transfer proposals]." The letter points out that in Kern County, as much as 50,000 acres of prime farmland could be fallowed, with 72,000 af/year of water no longer available for farming, as a result of shifts away from agricultural water use. Therefore, even though the DDWD–CLWA transfer accounts for only about one-fifth of that total, its environmental impact report should analyze the cumulative effect of this potential decline in irrigated acreage.

The JPA responded that although economic and social effects should not be treated as significant *environmental* effects, a cumulative analysis "could result in greater effects on the agricultural economy" (p. D-16). The total annual potential loss of revenue from all potential fallowing was $8.5 million (1984 dollars), roughly 0.1 percent of the value of the state's field, fruit, and nut crops. The potential loss of cotton amounts to 2.3 percent of California's total production, or 0.3 percent of national production. Further, since the price of water (paid to the State Water Project) was a significant farming cost at Devil's Den, the actual economic loss to the local economy would be less than if water originated there.

Just as there is little explanation as to why (besides expedience) the joint powers authority decided to include a cumulative impacts review in its final EIR, there is little explanation concerning the amendments the authority made to its environmental mitigation plans for Devil's Den Ranch as a result of the comments. One set of amendments involves changing the word *should* to *will* in the context of CLWA developing soil and water conservation plans, controlling weeds, and seeding the fallowed land to hasten a return to natural conditions once the farmland is permanently retired. This strengthens CLWA's commitment to the Devil's Den region following the water transfer.

The other interesting mitigation measure is a strengthened commitment to continue farming at Devil's Den Ranch. A new mitigation measure reads in part:

> [U]pon eventual reduction or elimination of the use of the State Water Project entitlement in Devil's Den, Castaic Lake Water Agency *would cause to continue an appropriate level of less-water-demanding agricultural activity* within the study area [italics added]. (p. 4-31)

This represents a significant shift away from the statement that farming may continue at Devil's Den Ranch to the extent it is economically feasible with the use of well water. Instead, CLWA was committing to farm "an appropriate level" of acreage even after taking the water. In the context of the final EIR discussion, determination of an appropriate level appears to include an examination of not only agricultural-market factors but also local economic conditions. This represents a strengthening of the economic commitment made by the receiving region to the region that was losing its surface-water supply.

CLWA's staff kept in close contact with the California Department of Water Resources throughout 1987 and 1988, regularly inquiring whether anything stood in the way of amending the SWP contracts. The answer regularly came back that the amendments were proceeding.

The final EIR is dated October 1988, and title to the land was transferred on October 25, 1988. Curiously, the SWP contract amendments were not finalized until a few days *after* CLWA had exercised its option on the land and closed escrow, which exposed it unnecessarily to a brief period of risk that the amendment process might come unraveled. It did not, and CLWA now has the right to transfer the water at a time and in increments of its choosing.

Little Change at Devil's Den; Economic Growth in the Santa Clarita Valley

In 1988, CLWA did not anticipate transferring the Devil's Den entitlement until the mid-1990s. When escrow closed, it immediately leased the ranch to a new farmer, who assumed the existing equipment leases and rehired nearly the entire Producers workforce. There was some confusion as to what equipment actually was there, however, and some theft resulted. The farmer's obligation to CLWA was based solely on water. CLWA charged $92/af for SWP water used on the farm, a price that was expected to cover CLWA's annual SWP costs for Devil's Den ($60–$65/af) and other costs of ownership.

A second farmer, Willie Rodrigues, took over the lease in 1992. Rodrigues employed his four sons on the ranch and cut back the workforce to about fifteen full-time-equivalent workers, well below the historical numbers of thirty to fifty employees. By 1996, none of the 1988 Producers employees was employed at Devil's Den; nearly all of them had moved on to new jobs or retired. About 6,000 acres per year were farmed for cotton, wheat, garbanzo beans, and melons. Rodrigues was taking 8,000–10,000 af/year of the SWP entitlement, depending on rainfall. His lease terms were the same as those for the previous farmer: based solely on his use of SWP water.

Through the years, CLWA has invested in the ranch, most significantly during the drought years of 1991–1992, when SWP agricultural deliveries were severely curtailed and a tumbleweed problem erupted on the fallowed fields. Tumbleweed eradication cost $350,000. There also have been efforts to eradicate morning glory growth as well as work evaluating the groundwater-well infrastructure.

With respect to environmental effects in the Santa Clarita Valley, CLWA's general manager, Bob Sagehorn, believes that acquisition of the Devil's Den entitlement has exerted a "stabilizing influence" on development activities. Even without transferring water, CLWA has been able to certify the availability of a long-term water supply, and this has enabled developers to proceed with residential and commercial building projects.

The transfer has not escaped high-profile criticism, however. An article on California's water-supply challenges in a 1993 special edition of the *National Geographic* takes the entire transfer to task. It asserts that "Devil's Den had succeeded in selling out a few years back" so that Santa Clarita could "make itself into the next San Fernando Valley" (Conniff 1993, pp. 50–51). The CLWA service area is then described:

> Clusters of pink-roofed new housing spilled over the ridgelines and huddled in the ravines. Heavy machinery was regrading the rumpled hillsides. (p. 51)

Later in the article, a local official explains that in the Santa Clarita Valley, water, politics, and real estate money were still tied as closely as they were during the Owens Valley era. One major developer, the Newhall Land and Farming Company, also owned the second largest private water company, through which it pressured CLWA to acquire more water rights. The local official links the acquisition of Devil's Den water with a proposed 1,800-unit housing-project-cum-golf-course whose site was designated a "significant ecological area" (Conniff 1993, p. 52). A CLWA official responds to these criticisms by noting that the water district's role is to provide water, not to set development policy, which is the job of city and county planners.

The criticisms of the *National Geographic* focus more on the process and form of development in the Santa Clarita Valley and less on the acknowledged underlying pressure of continued population growth. The 1995 population of CLWA's service territory was 150,000, less than what had been projected in the mid-1970s. Two separate 1995 growth projections suggest 2010 populations of 270,000 and 357,000. CLWA planning documents project a water demand of 120,000 af/year to serve a population of

270,000. This figure includes both commercial services for the growing population and an assumption of shrinking numbers of persons per household. Although the demand figure suggests generous allotments per household, even if it is off by a factor of two it would still exceed CLWA's post–Devil's Den entitlement of 53,900 af/year. Thus, although valid criticism may be leveled at the planning process (especially one that projects declining numbers of persons per household—and therefore a need for more houses for the same population—when a housing developer heavily influences planning), the underlying demographics suggest that CLWA's effort to acquire additional water rights was not frivolous. By 1998, CLWA had not yet taken delivery of its Devil's Den water.

Conclusion: More Lessons for Statewide Transfer Institutions

Of the three recent long-term water transfers in California, the DDWD–CLWA deal provides the only case of "vertical integration": instead of trading with a legally separate party, the buyer and seller joined up to form a single entity and then undertook an internal transfer of the water entitlement. This contracting choice fundamentally altered the organization of the agreement.

CLWA faced potential contracting risks. It was investing in farmland in order to acquire water. If the water ultimately was not transferable, CLWA could have been stuck with land whose value no longer included the value of a potentially transferable water resource. By fashioning an option agreement, CLWA created a security feature that allowed it to undertake the transfer effort before buying the land and then to purchase the land when the transfer was ensured. The option-to-purchase contract mitigated CLWA's contracting risk.

For Producers, there was little risk beyond the time and capital invested in negotiations with CLWA.

The simplicity of the DDWD–CLWA agreement might suggest that other regions, rural and urban, could make similar agreements. But several unique features facilitated this water transfer. First, both Devil's Den Ranch and DDWD were owned and controlled by one entity: Producers Cotton Oil Company. All members of the district's board were employees of Producers. Only one decision maker, with one set of interests, existed on the selling side.

Second, the historical irrigation infrastructure, which in the 1980s was described as "all earthen canals and gopher holes" (Hull 1996), made the property less attractive for farming but more attractive in terms of a water transfer. The price of the ranch would not have to include the value of the irrigation system. Also, had any local groups opposed the transfer, they would not have been able to point to the existence of a state-of-the-art irrigation system rusting away while local agricultural water moved to the city.

Third, the aquifer beneath Devil's Den Ranch was hydrologically separate from other aquifers, which meant that a reversion to well farming would not harm nearby farm operations.

Fourth, Devil's Den Ranch is located in an extremely isolated area. Rarely used Highway 33 travels north to south through the ranch. No town is closer than eighteen miles away, and the ranch had a resident population of roughly twenty people. The region's only environmental notoriety arose from the presence of the endangered San Joaquin kit fox, on which the transfer probably would have had minimal effect. With this kind of isolation, it is not surprising that no challenge was made to the SWP's contract amendment process and only three substantive comment letters were submitted on the draft EIR.

Fifth, Producers was a highly motivated seller. The new owner of Producers, Dunavant Enterprises, Inc., had a policy of *not* owning cotton-producing farmland, and the company was large enough that expedience could outweigh the balance-sheet effects of a below-market sale of a parcel as small as Devil's Den Ranch. Further, Devil's Den Ranch was the last remaining parcel Producers owned after selling off its larger pieces. Thus, any buyer was likely to get a good price.

Finally, imported water might never have come to Devil's Den had the coastal branch of the California Aqueduct not been slated to pass through the ranch property. This reduced significantly the cost of delivering water from the main aqueduct to Devil's Den Ranch, increasing the financial feasibility of its becoming a SWP contractor in the first place. Since both the buyer and the seller were SWP contractors and took water from the California Aqueduct, and the water would have already passed through the Sacramento–San Joaquin Delta no matter where it was applied, transaction-specific conveyance problems were all but nonexistent.

For all these unique elements that helped enable transfer of the Devil's Den entitlement, the agreement still took more than two years to finalize from first meeting to signed contract, and another year was needed to secure the water transfer. The parties not only signed the option agreement for the land purchase but also made a side agreement on creating the joint powers authority to prepare the environmental impact report. As part of getting the water transferred, the agreement created a long-term supportive relationship between the suburban water district and the rural area that had given up its surface-water entitlement.

CLWA fared well with this transfer. Under essentially the same terms as its existing SWP contract, it had augmented its long-term supply of water. Until the water was needed, CLWA could sell it to its Devil's Den Ranch lessee at a price that covered the cost of the water and even bit into the ranch's original sales price. The water could then be transferred incrementally as it was needed in the Santa Clarita Valley. The only contractual difference at the time was that the Devil's Den water retained its agricultural classification. However, the 1994 Monterey Agreement, which is described in appendix 1, did away with the agricultural/urban classification system, so CLWA no longer has to worry about the ultimate reliability of the Devil's Den water after it is transferred. One CLWA official commented that California may never again see a rural-to-urban transfer of water at a price so low and ultimate delivery terms so flexible.

With respect to historical precedent, this agreement was structured almost exactly the same way as the agreement that secured rights to the Owens River for the Los Angeles Department of Water and Power (LADWP) at the beginning of the twentieth century. Both transfers were examples of vertical integration: LADWP bought options from Owens Valley farmers to purchase their acreage, exercised the options, and then transferred the water rights to enable growth of the San Fernando Valley. In both cases, an urban or suburban water district became a rural landowner as a result of water acquisition. Since any transfer that takes the same contractual form as the Owens Valley transfer raises appropriateness concerns, a brief comparison of the two situations is in order.

There are key differences. Owens Valley certainly was an "area of origin" for Owens River water. Owens Valley farmers, as appropriators, were water-rights holders. DDWD was simply a contractor for SWP water and held an SWP entitlement. The area of origin for Devil's Den water is probably the Feather River watershed in the northern Sierra Nevada. Further, SWP water had been used at Devil's Den Ranch only from 1970 to 1988, preceded by thirty-five years of groundwater farming. Also, although the DDWD–CLWA negotiations themselves were conducted privately and quietly, the EIR process allowed the public to review and critique the proposal. In contrast, LADWP's negotiating record is one of secrecy and deceit. Finally, historically there has been almost no economic activity anywhere near Devil's Den Ranch, and employment at the ranch itself was in a multiyear decline. This again is unlike the situation in the Owens Valley, which had a bright economic future at the dawn of the twentieth century. The lesson that emerges from this comparison is that future land-purchase agreements should be judged on their own merits and not simply dismissed because of similarities to a dark episode in California's water history.

Notes

1. Two generations after the subdivision, Producers tried to track down owners of the lots in order to repurchase them and simplify district ownership, but the effort turned into a nightmare because many descendants had migrated to other countries, had simply disappeared, or held the plots jointly with other scattered family members. Ultimately, Producers turned over the farming leases for some of these plots to the Castaic Lake Water Agency, with actual ownership either unknown or too confusing to sort out.
2. The Williamson Act taxes land dedicated to farming at a lower rate than land that can be converted to other uses, such as suburban growth. If Williamson Act land is reclassified, its owners may have to pay extensive back taxes.

CHAPTER 8

Getting to a Market: Lessons and Design Recommendations

"Successful water transfers do not emerge from thin air."
—Smith and Vaughan 1993

This chapter distills insights and lessons from California's experience that should be considered in future efforts to reallocate water in California and other western states. It then presents twelve recommendations that I believe should guide future efforts to reallocate water from rural to urban areas. The recommendations are intended to provide California with a template for reform that is feasible and appropriate, and one that I believe is preferable to both the current policy logjam and the leading pro-market alternatives. The chapter then notes the importance of rural–urban cooperation regarding water and other shared interests in the coming decades and concludes with a discussion of the possible future role of water marketing in reallocation plans in the western United States.

Why Change California's Water Policy, and Why Change to a Market?

Throughout the twentieth century, California's water policy moved away from strong private-property rights and toward greater public participation in decisions about water use and water allocation. Areas of origin and environmental interests are among the primary beneficiaries of this evolution of policy. The evolution has taken place in legislative, administrative, and judicial arenas and across a broad spectrum of water issues. Today's efforts to create a statewide water market run contrary to the twentieth century's evolution of property rights in water.

Another trend until recent decades was to develop new water supplies to meet new demand. It is likely that the era of dam building in California has drawn to a close, although engineering projects that increase flexibility of use, such as construction of new storage facilities, are likely to be undertaken. From development of new supplies, state and local water policy has shifted to conservation, reuse, and adaptation to shrinking water budgets relative to demand. The most recent policy thrust has been institutional reform to enable reallocation of water rights, including reallocation through market-based transfers. As chronicled in appendix 1, great public energy over the past two decades has gone into creating water markets, but few results have been realized in terms of long-term interregional rural-to-urban transfers.

There are valid reasons to pursue a policy of water reallocation. They include demographic projections of expanded population growth in California's coastal (as well as inland) cities, the need for reliable water supplies to fuel industrial growth, and widely held values favoring preservation and restoration of ecosystems and endangered species. A reallocation policy involving a market would provide the attractive benefits of both promoting economically efficient outcomes and giving holders of water rights greater flexibility in ownership. A market also could form the basis of a political compromise when other attempts at creating a reallocation policy have failed.

An effective market mechanism could both preserve hard-won public benefits related to water allocation and avoid risks associated with decentralized, incremental management of a public resource. There are risks not only on the water-selling side but also on the water-buying side, as described extensively in this book. The rapid transformation of open land and farmland into suburbs is one example of a land-use issue that is closely connected to water reallocation choices.

Lessons from California's Short-Term Markets and Long-Term Transfers

But how does one move from the hypothesized potential of a water-market mechanism to an actual mechanism? And how does one secure the benefits promised by a market mechanism while avoiding its pitfalls? This section presents insights and lessons from California's experience with both short- and long-term transfers. The goal is to integrate the state's historical experience with water rights and water transfers with the inherent strengths and weaknesses of markets in order to create a set of feasible and appropriate institutions for California.

The state's experience with short-term water transfers provides a number of insights into the formation of long-term markets. First, short-term transfers have not threatened the state's agricultural sector with permanent reductions of agricultural water rights. Many farmers and irrigation districts make a distinction between water as a *commodity*, and therefore able to be sold in bulk to the highest bidder, and water as an *input to the agricultural economy*. The scope of a short-term market is well defined and non-threatening, whereas the potential quantities of water involved in an open rural-to-urban market could be a threat to agricultural ways of life. The formation of any long-term rural-to-urban market will have to address issues of market scope as well as direction of reallocation. This discussion will inevitably touch on the historical context of Owens Valley. It should therefore be open, participative, and credible to the agricultural community. Proposed reforms that do not address the scope issue will probably be too threatening to the agricultural sector to be adopted. In other words, policy makers cannot simply form a market and then "let the market decide" how much water is reallocated and where it is reallocated. In order to create an effective market mechanism, decisions about the intended reallocation results of the mechanism must first be made. With a clear policy goal in mind, the market becomes a tool of policy implementation, not a goal in and of itself.

Other lessons from short-term markets include the importance of reliable and low-cost conveyance, minimal adverse effects on the environment and on third parties, and transactions whose costs and benefits are easy to calculate and whose terms are easy to negotiate, conclude, and execute. With respect to third parties, not only should direct adverse effects be avoided, but also trading programs should strive for fairness in terms of access to markets and opportunity to trade.

The three long-term interregional rural-to-urban transfers detailed in chapters 5–7 are striking in both their similarities and their differences. They are similar in that all were intended to move water toward California's southern coastal region and involved water "wholesalers" as buyers. Each took years to negotiate and required more than one formal agreement.

Another similarity is that no water *rights* actually changed hands. The Palo Verde Irrigation District (PVID) and the Imperial Irrigation District

(IID) simply took (and still take, in IID's case) less of their share of Colorado River water, leaving more for the Metropolitan Water District of Southern California (MWD) under its lower priority. With the Devil's Den Water District (DDWD), a contractual entitlement with the State Water Project (SWP) was the object of transfer. The state Department of Water Resources (DWR) is the holder of water-rights; its status was not changed by the transfer agreement.

None of the agreements has resulted in significant negative effects on third parties in rural areas. PVID's labor effects were minimized by a number of factors; IID's conservation efforts created construction jobs but appear not to have affected farm employment; and DDWD's combination of isolation and continued irrigation has rendered negligible any local adverse effects of that transfer. The lack of adverse effects on labor here does not mean, however, that no rural-to-urban transfer will affect labor. An equally valid conclusion is that one reason these three are the only transfers to have been concluded is that their labor effects were negligible. Labor effects should be addressed both in the design of water-market institutions and in the context of future individual transfers.

In no case were third parties that did not also hold legal water rights included in the design or negotiation of the agreements, except explicitly to exclude them. In all three cases, however, there was feedback from interested parties who did not hold legal water rights. These parties were not at the negotiating table, but they could attend town meetings (as in PVID), write to local newspapers and vote for or against irrigation-district board members (IID), and comment on the proposed environmental impact reports (DDWD). Feedback was limited, however, to those parties who were aware of the transfer negotiations and agreements and were organized and ready to act when the opportunity came. In the case of DDWD, it was other governmental offices representing Kern County and the state that commented on the draft EIR, not environmental or community-based nongovernmental organizations. These transfers' level of formal exclusion was a negotiators' luxury not likely to be repeated in future rural-to-urban transfers.

Despite the importance of considering third-party effects when transferring water, it is misleading to think of rural towns as passive future victims of rural-to-urban water transfers. Rural towns may not be (and need not be) as dependent on and vulnerable to water-use decisions of surrounding farmers and districts as is popularly believed. Business leaders in Blythe, the largest city in the Palo Verde Valley, saw that agriculture-related

APPENDIX 3

Proposals for Market-Based Reallocation: Evaluating Their Feasibility and Appropriateness

"Human beings owe each other help to distinguish the better from the worse."

—John Stuart Mill, *On Liberty* (1859)

Most evaluations of market mechanisms center on the familiar dichotomy of efficiency versus equity. An efficient allocation results in all goods ending up in the hands of those who value them most, given everyone's initial individual wealth. If one party can make better use of a good than the party who currently owns it, the two parties will strike a deal and transfer the good. Efficient markets reduce waste because all goods either are put to their best economic use by the current owner or are sold to somebody else who will do so. An economically efficient system attaches no preference to the direction of trading (i.e., who and where are the buyers and who and where are the sellers). Traders are faceless; what matters is that if potential gains are to be had from trade, the trade is done.

Equity refers to a society's sense of fairness with respect to price-based allocation of goods and services. On the positive side, the vision of a nondiscriminatory choice of trading partners has been a leading moral argument in favor of pro-market reforms for two centuries. The market mechanism, with its promises of fairness and general prosperity for all participants, draws moral validity from its perceived neutrality with respect to who ends up with money, who ends up with goods, and in what combinations.

But from an equity perspective, markets also have downsides. When price is the arbiter of who gets what, the rich have an advantage over the poor. For some goods and services, such as provision of potable water for human use, pure market-based allocation assaults one's sense of equity. With a "pure" market, the poorest in a society would probably be priced out of potable-water use, with resulting negative effects on human health and quality of life. Even in bulk purchases of untreated water, poorer regions might suffer setbacks if their water is transferred away. Thus, the typical discussion of markets versus other forms of resource allocation pits the relative superiority of a market in terms of its efficiency against the relative superiority of the proposed alternative in terms of its equity.

Equity and efficiency are not the only standards by which market mechanisms can be judged. In fact, the two are similar in a critical sense: both are concerned with how appropriate a market mechanism is for a given task in a given context. That is, both arguments are structured in the same way: "This proposed market mechanism is efficient or equitable, and that is appropriate." The broader underlying standards for evaluating the effectiveness of a reallocation mechanism can therefore be labeled *appropriateness*. A second, similar standard can be labeled *feasibility*. Efficiency and equity, along with other considerations, would fall under the appropriateness standard; other considerations would fall under the feasibility standard.

Feasibility refers to the instrumental nature of the proposal: to what extent does it achieve the policy goal that motivates its consideration? In the case of water reallocation in California, the policy goal is to shift water use from agricultural applications to urban and environmental applications. The feasibility standard further includes the question of implementation: can the plan, once approved, actually be installed and made operational? One aspect of feasibility not covered concerns the matter of securing the political go-ahead (e.g., legislative votes or key signatures) to launch the program. This is because an analysis of a program's feasibility would generally precede and inform the political process that selects the program.

Appropriateness refers to the proposed policy from the perspective of social values. To what extent is the proposal, including its implementation mechanism, consistent with a society's values, narrowly or broadly construed? Different lists of criteria would be generated on the basis of the attributes of the society, the resource in question, and the reallocation goals. Equity, efficiency, respect for the individual, respect for communities, respect for nature, reduction of wasteful use, and protection from disasters are among the long list of sometimes contradictory values that should be considered in the context of reallocating California's water.

The appropriateness standard matters not just as a subject for academic conversation: the California state legislature itself has called for water transfers that are consistent with the well-being of both sending and receiving regions as well as with protection of nature. That is, the legislature wants transfers that are appropriate for the state given its current needs and values. (The quote is attached to note 2 on page 154.)

The categories of feasibility and appropriateness are not completely distinct. That is, it is possible that a proposal could be infeasible because it is inappropriate. Parties who perceive that they would be unfairly treated by a proposal may have the legal or administrative means to block its implementation. So although feasibility and appropriateness are separate standards in terms of their focus (instrumentality versus consistency with values), they also can be mutually reinforcing.

Figure A.3.1 is a matrix with five boxes in which separate evaluations of a market mechanism (or other policy implementation option) could be made. On the left of the matrix are two standards of comparison, feasibility and appropriateness. Along the top of the matrix are three modes of comparison:

1. How does the proposal compare with other proposals intended to accomplish the same goal?

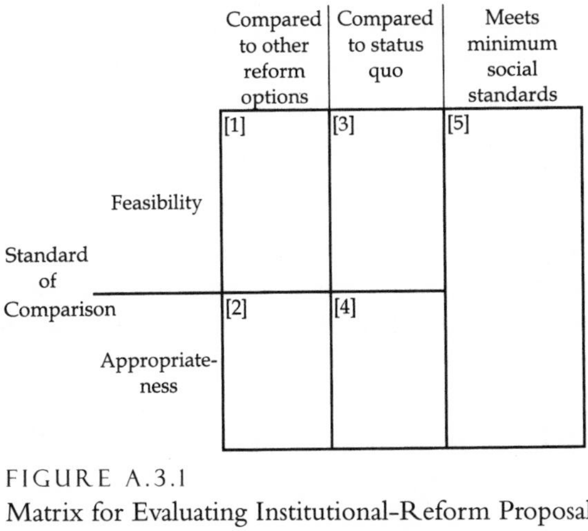

	Compared to other reform options	Compared to status quo	Meets minimum social standards
Feasibility	[1]	[3]	[5]
Appropriateness	[2]	[4]	

Standard of Comparison

FIGURE A.3.1
Matrix for Evaluating Institutional-Reform Proposals

2. How does the proposal compare with the existing set of institutions, or the status quo? Existing ways of doing things have the built-in advantage of having had practical experience and having adapted over time to specific, complex circumstances that might not be apparent to program designers.

3. Does the proposal meet minimum social standards? Even if it is shown to be superior to all other alternatives and the status quo, the proposal still may not be "good enough." If pursuing a not-good-enough option forecloses the opportunity to implement a future, superior option (perhaps because of a fixed budget or an irreversible adverse effect associated with implementation), a decision maker may be better off not adopting the otherwise superior existing option and waiting for a better one to be developed. A proposal to build a major new dam on an ecologically significant waterway might fall into this category.

Each of the boxes in figure A.3.1 asks for a separate evaluation. Boxes 1 and 3 are concerned with comparative feasibility tests. For example, one can test for comparative feasibility of a water-market proposal by examining how costly it would be to transact. How many legal steps need to be taken to complete the deal? How costly are the steps? How time-consuming are the steps? Who is responsible for enforcing the agreement? What enforcement mechanisms are available? How flexible are the transfer rules in terms of unexpected future contingencies?

An analysis of boxes 2 and 4, which are concerned with comparative appropriateness, might include an examination of the extent to which proposals could injure income equality, access to future economic opportunity, sustainable development, or biodiversity. The extent to which a proposal reduces existing wasteful uses of water, as well as other aspects of the efficiency question, could also be explored. One might further examine whether and to what extent proposals are consistent with long-term trends in legal interpretation or precedents, which could stand as a proxy for a society's evolving cultural relationship with water.

An analysis pertinent to box 5 could include an examination of comparative costs for achieving other similar policy goals, potential adverse effects on public health, or irreversible adverse environmental effects such as species extinctions or landscape reconfiguration.

The value of this matrix is that it elaborates and clarifies the scope and comparative nature of the proposal-evaluation process. It also sorts out and makes space for both normative and positive criteria, both of which are necessary for a thorough evaluation. The boundaries around each box are not impermeable. Indeed, the discussion of the evolution of property rights in chapter 3 contains perspectives that would fall into all three appropriateness boxes (2, 4, and 5).

The use of figure A.3.1 is not likely to yield an unequivocal, incontestable recommendation. Simply in terms of a binary better-or-worse ranking, there are thirty-two possible outcomes for any proposal, ranging from it being consistently superior to a single alternative proposal and the status quo to its being consistently inferior. Certainly, the closer a proposal is to arriving at a consistently superior or a consistently inferior result, the easier it is to justify a recommendation consistent with the result. Setting aside extreme results, one would also have to take into account the degree of superiority or inferiority of the various comparative categories. If a proposal is found to be substantially inferior to the status quo in terms of either appropriateness or feasibility, that could stand as grounds to cease further analysis and move on to an analysis of other existing proposals or formulation of new ones. Once formed, new options can then be subjected to the same rigorous analysis.

Farmer Resistance to Long-Term Rural-to-Urban Water Transfers: Two Examples and an Economic Interpretation

> *Item, 1984:* Two representatives of a suburban water district located on California's rapidly expanding southern coast identify a farmer in the upper Owens Valley who wishes to sell his water rights. Conveyance of the water can be arranged through the Los Angeles Aqueduct. The representatives travel to the Owens Valley for a public hearing on the proposed transfer. On arrival at the meeting-site parking lot, the officials realize that nearly every parking space is taken by pickup trucks and all the gun racks are empty. They get back in their car and drive home.

In recent years, there has been intense social pressure in the agricultural sector against any farmer selling water to urban regions. The following are two examples. First, in 1993, the Central Valley's Areias Dairy Farm (ADF), facing bankruptcy, attempted to become the first farm to sell its water rights according to the rules of the 1992 Central Valley Project Improvement Act. The agreement reportedly called for the Metropolitan Water District of Southern California (MWD) to pay ADF $175 per acre-foot (af) for about 35,000 af of water over a period of fifteen years. MWD would be able to purchase as much as 5,000 af/year in each of any seven years it chose; ADF would receive as much as $6.1 million. Other reports put the quantity transferred at a slightly smaller amount. Two public hearings in August 1994 near the farm brought out well more than 1,000 community members, nearly all of whom opposed the trade, citing its potential adverse effects on the local community and economy. The Merced County Farm Bureau also came out against the proposed transfer, as did the area's irrigation district, the Central California Irrigation District, representatives of which said they wished they had been consulted in advance of the agreement. ADF and MWD battled for months to implement the agreement. A commissioned study pointed out that ADF occupied only 1 percent of the total land in its irrigation district, and the transferred water represented only 0.75 percent of the district's total normal-year allocation. The farm's production contributed only 0.5 percent of the county's economic output. However, opposition from the farm sector was enough not only to scuttle the trade (with the farm eventually declaring bankruptcy) but also to deal a temporary blow to the political career of Rusty Areias, a son of the farm's owner who served as state assembly representative from nearby Salinas, a city devoted to the farming industry. Areias stood by his family's right to transfer water rights. Although many other issues weighed in during the subsequent 1996 campaign, Areias eventually lost a tight race.

The second recent example of local opposition to a long-term rural-to-urban transfer involves a water-storage agreement. In 1997, a rejuvenated effort by MWD and the Arvin-Edison Water Storage District to reach a long-term agreement to move San Joaquin River water to the southern coast encountered intense opposition from Central Valley farmers. Farmers insisted that not one drop of San Joaquin River water be transferred to urban use in southern California because that could be the first step in a trend that would threaten all agricultural users of the river.

Both of these recent cases of opposition by farmers to a water transfer are consistent with a simple economic model of farmers' attitudes toward water transfers. The following paragraphs provide at best only a partial explanation for the farmers' opposition but may shed some light on their behavior with respect to water markets.

For nearly two decades, California's odd institutional setting has included both a *lack* of a long-term market in terms of actual transactions and a perception that there *is* or *could soon be* a market, in terms of institutional structures. Today's "starting place" for a long-term water market, should one arise, finds agricultural water districts in control of about 80 percent of developed water and cities with about 15 percent.[1] Cities are demanding water and can afford to pay handsomely for it. Today's lack of a water market has created a supply bottleneck, and as a result, purchase prices offered by cities are high: well over $100/af. Indeed, permanent transfers of State Water Project (SWP) entitlements were reportedly selling for $1,000/af in late 1997, with subsequent annual payments of $60 to $100/af delivered added to the agreement.

But with so much *potential supply* available to a market, should one arise, it is likely that once market transfers begin, the actual market-clearing price will be much lower than what is now being quoted. In 1995, an MWD employee in the water-marketing group privately described the agency's buying strategy as one of offering high prices at first to bring the first few farmers to the table and close some deals and then lowering the offered price significantly as trading takes off and offers to sell come pouring in.

Against this backdrop, consider one possible strategy of an individual farmer. He or she wants either to be the first to sell or to have no one else ever sell. Even if just a few other farmers sell water to cities first, the value of his or her water rights could drop precipitously—perhaps by more than 50 percent. A simple and conservative calculation suggests the economic risk to the state's farm economy if a handful of long-term rural-to-urban transfers actually take place. Assume that the value of a farmer's water rights is part of the capital value of the farm and that roughly 10 million acre-feet (maf) of water is *available* for rural-to-urban transfer, even though nowhere near that amount could ever actually be transferred. If, with the advent of a water market, the market-clearing price (excluding conveyance) for permanent water rights drops by $500/af (say, from $1,000/af to $500/af), then the perceived capitalized value of the farming sector in regions with transferable water rights could drop by $5 billion in a matter of weeks. At a minimum, this loss of value would erode the rural sector's creditworthiness, making borrowing more costly and requiring higher profit margins. This in turn would make farming in California less competitive with that in other farming regions. From this perspective, the advent of long-term water marketing presents an economic threat to large portions of California's rural economy.

If a long-term market is to arise, either it must be organized in a way that does not threaten the capital value of the state's farming sector due to the potential statewide drop in the value of agricultural water, it must be organized in a way that provides a substitute for the lost value, or it must be phased in in a way that allows farmers and rural regions to adapt effectively to the loss of capital value. The latter will most likely occur.

Note

1. The remaining 5 percent is used by federal and state governments, Indian tribes, and others.

Source Materials

Much of this book is based on my doctoral dissertation (Haddad 1996). The dissertation, as one might expect, is thick with references. To enhance the flow and readability of this book, I pared the references down considerably. If scholars or other parties are interested in pursuing further the points made here, they may peruse the dissertation, a document that is more detailed but less user-friendly. The dissertation is available at the main library of the University of California, Berkeley.

Adams, Rev. Bobby. 1988. Advertisement. *Imperial Valley Press,* December 25.

Arax, Mark. 1995. "Effort to Link Growth, Water Sparks Battle." *Los Angeles Times,* August 14.

Bakken, Gordon Morris. 1983. *The Development of Law on the Rocky Mountain Frontier.* Westport, Conn.: Greenwood Press.

Bay Area Economic Forum. 1991. *Using Water Better: A Market-Based Approach to California's Water Crisis.* October. San Francisco: Bay Area Economic Forum.

Becker, Lawrence C. 1977. *Property Rights: Philosophic Foundations.* London: Routledge & Kegan Paul.

Boronkay, Carl, and Timothy Quinn. 1993. "The Central Valley Project Improvement Act: An Urban Perspective." *San Joaquin Agricultural Law Review* 3:57–64.

Boronkay, Carl, and Charles Shreves. 1989. "Water Trades Can Help Meet Future Urban Needs." *EDF Letter* 20, No. 2 (May): 7.

Calexico Chronicle. 1989. "Secretary Hodel Applauds California Water Conservation Agreement." *Calexico Chronicle,* January 26.

Chen, Ingfei. 1991. "Water Department Gets OK to Buy on Open Market." *San Francisco Chronicle,* January 30, A-3.

Coase, Ronald. 1937. "The Nature of the Firm," *Economica N.S.* 4:386–405. In *Readings in Price Theory,* ed. G. J. Stigler and K. E. Boulding. Homewood, Ill.: Richard D. Irwin.

Conniff, Richard. 1993. "California: Desert in Disguise." *Water: The Power, Promise, and Turmoil of North America's Fresh Water,* National Geographic Magazine Special Edition, pp. 38–53.

Cooter, Robert, and Thomas Ulen. 1996. *Law and Economics.* Reading, Mass.: Addison-Wesley.

Cox, Don. 1989. "Positive Move." Editorial. *Imperial Valley Press,* January 17.

Daly, Herman E. 1996. *Beyond Growth: The Economics of Sustainable Development.* Boston: Beacon Press.

Devil's Den–Castaic Lake Authority. 1988. *Environmental Effects Outside the Castaic Lake Water Agency Service Area Resulting from the Potential Transfer of All or Part of the Contractual Water Entitlement of Devil's Den Water District to the Castaic Lake Water Agency.* Final Environmental Impact Report. October. State Clearinghouse No. 88042504. Newhall, Calif.

Dimmitt, Arnold K. 1994. "The Imperial Irrigation District/Metropolitan Water District Water Conservation Program: A Case Study." Unpublished manuscript.

Dixon, Lloyd S., Nancy Y. Moore, and Susan W. Schecter. 1993. "California's 1991 Drought Water Bank: Economic Impacts in the Selling Regions." Report prepared by the Rand Corporation for the California Department of Water Resources.

DWR (California Department of Water Resources). 1986. *A Catalog of Water Transfer Proposals.* Draft. September. Sacramento, Calif.: DWR, Water Transfers Committee.

———. 1990. *A Catalog of Water Transfer Proposals—an Update.* Draft. December. Sacramento, Calif.: DWR, Division of Local Assistance.

———. 1992. *Weekly Update on Drought Conditions.* March 13. Sacramento, Calif.: DWR.

———. 1998a. *California Water Plan Update: Bulletin 160-98.* Executive summary. Sacramento, Calif.: DWR.

———. 1998b. *California Water Plan Update: Bulletin 160-98.* Vol. 1. Sacramento, Calif.: DWR.

Elias, Tom. 1995. "Marketing Idea Key to Ending Water Wars." *Torrance Daily Breeze.* March 6, B-2.

Environmental Defense Fund. 1989. "California Agencies Sign Historic Water Swap." *EDF Letter* 20, No. 2 (May): p. 1.

Fallowing Agreement. 1992. Agreement for Land Fallowing in the Palo Verde Irrigation District ("Fallowing Agreement"). Agreement between the Metropolitan Water District of Southern California and the Palo Verde Irrigation District.

Gardner, B. Delworth. 1987. "Removing Impediments to Water Markets." *Journal of Soil and Water Conservation* (November–December): 384–388.

Garner, Eric L., Michelle Ouellette, and Richard L. Sharff Jr. 1994. "Institutional Reforms in California Groundwater Law." *Pacific Law Journal* 25:1021–1052.

Gleick, Peter H., Penn Loh, Santos V. Gomez, and Jason Morrison. 1995. *California Water 2020: A Sustainable Vision.* May. Oakland, Calif.: Pacific Institute for Studies in Development, Environment, and Security.

Great Western Research, Inc. 1995. "Palo Verde Test Land Fallowing Program, August 1, 1992–July 31, 1994, Final Report." Vol. 1. August. Report prepared for the Metropolitan Water District of Southern California.

Griffin, Ronald C., and John R. Ellis. 1997. "Water Marketing in Texas." Draft. July 15. College Station: Texas A&M University, Department of Agricultural Economics.

Grover, George S., and John F. Mann Jr. 1991. "*Acton v. Blundell* Revisited: 'Property' in California Groundwater." *Western State University Law Review* 18:589–598.

Haddad, Brent M. 1996. "Evaluating the Market Niche: Why Long-Term Rural-to-Urban Inter-regional Markets for Water Have Not Formed in California." Ph.D. diss., University of California, Berkeley.

———. 1997. *Putting Markets to Work: The Design and Use of Marketable Permits and Obligations.* Working Paper No. 19. Paris: Organization for Economic Cooperation and Development, Service de la Gestion Publique.

Hayek, F. A. 1945. "The Use of Knowledge in Society." *American Economic Review* 35, No. 4:519–530.

Hof, Robert D. 1992. "California's Next Cash Crop May Soon Be . . . Water?" *Business Week* (March 2): 76–78.

Honoré, A. M. 1961. "Ownership." Pp. 107–147 in *Oxford Essays in Jurisprudence: A Collaborative Work,* ed. A. G. Guest. London: Oxford University Press.

Howitt, Richard, Nancy Moore, and Rodney T. Smith. 1992. "A Retrospective on California's 1991 Emergency Drought Water Bank." March. Report prepared for the California Department of Water Resources.

Hull, Robert. 1996. Farm manager, Devil's Den Water District, retired. Personal interview, Corcoran, California, February 13.

Hyduke, Jeanette. 1991. "MWD Seeks Farm Land for Water." *Riverside Press-Enterprise,* December 7.

Kamand, Fadi. 1995. Engineer, Metropolitan Water District of Southern California. Personal interview, Los Angeles, March 6.

Kaplan, Karen. 1991. "State Starts Water-System Reform: Five-Year Drought Has Demonstrated Marketing Is the Way to Go." *San Diego Union,* July 21.

Kay, Jane. 1994. "Big Businesses Want Secure Water Supply." *San Francisco Examiner,* July 21.

Lapin, Lisa. 1991. "'Perestroika for Water': Drought Inspires Revolution." *San Jose Mercury News,* February 3.

MacDonald, Clyde. 1993. "Water Supply: A New Era for a Scarce Resource." Chap. 6 in *California's Threatened Environment,* ed. Tim Palmer. Washington. D.C.: Island Press.

Morris, Willy. 1989. "Water Swap Finalized After Five Years of Talks." *Imperial Valley Press,* January 1.

MWD (Metropolitan Water District of Southern California). 1986. "Metropolitan's Detailed Comments on the Draft Environmental Impact Report—Proposed Imperial Irrigation District Water Conservation Program and Initial Water

Transfer." June 13. Reprinted in *Final Environmental Impact Report*, October 1986, 4–87 ff. Los Angeles: MWD.

———. 1989. *Water Conservation Agreement between the Metropolitan Water District of Southern California and the Imperial Irrigation District*. Contains copies of the Water Conservation Agreement, Approval Agreement, and Agreement to Supplement Approval Agreement. December. Los Angeles: MWD.

———. 1990. *The Regional Urban Water Management Plan for the Metropolitan Water District of Southern California*. November. Los Angeles: MWD.

———. 1994. "Integrated Resource Plan: Phase 1 Report." Final draft. November 28. Los Angeles: MWD.

Olmstead, Janis. 1997. "Emerging Markets in Water: A Comparative Institutional Analysis of the Central Valley Project and the Colorado–Big Thompson." Working paper. September 22. Berkeley: University of California, Department of Agricultural and Resource Economics.

Porgens, Patrick, and Lloyd G. Carter. 1992. "How to Get Rich off Public Water." *Sacramento Bee*, February 16.

Producers Cotton Oil Company. n.d. Company brochure, circa 1987. Memphis, Tenn.: Producers Cotton Oil Company.

Program Agreement. 1992. "Agreement for the Implementation of a Test Land Fallowing Program and Use of Saved Water (Program Agreement)." Agreement between the Metropolitan Water District of Southern California and the Palo Verde Irrigation District. May 29.

PVID (Palo Verde Irrigation District). 1994. "Palo Verde Irrigation District, Blythe, California." Informational handout. August. Blythe, Calif.: PVID.

Reinhold, Robert. 1992. "Farmers in West May Sell Something More Valuable Than Any Crop: Water." *New York Times*, April 6.

Rice, P. A. 1990. "Letter Urged Telling Sea's Problems in Conservation." *Imperial Valley Press*, May 25, A-1.

Robie, Ronald, and Russel Kletzing. 1979. "Area of Origin Statutes—the California Experience." *Idaho Law Review* 15:419–441.

Rosegrant, Mark W. 1995. "Water Transfers in California: Potentials and Constraints." *Water International* 20:72–87.

Rosenberg, Richard M. 1995. "A New Era in California Water: The Business Perspective." Speech delivered to the Water Education Foundation, March 30. Sacramento, Calif.

Sagehorn, Robert. 1995. General manager, Castaic Lake Water Agency. Personal interview, Castaic Lake, California, June 12.

San Francisco Chronicle. 1991. "Emergency Water Bank Has a Surplus." *San Francisco Chronicle*, November 19.

Sax, Joseph L. 1990. "The Constitution, Property Rights, and the Future of Water Law." *University of Colorado Law Review* 61:257–282.

Sax, Joseph L., Robert H. Abrams, and Barton H. Thompson Jr. 1991. *Legal Control of Water Resources*. 2nd ed. St. Paul, Minn.: West.

Schulz, Clifford W., and Gregory S. Weber. 1988. "Changing Judicial Attitudes towards Property Rights in California Water Resources: From Vested Rights to Utilitarian Reallocations." *Pacific Law Journal* 19:1031–1110.

Severns, Marjorie. 1989. "Voice of the People." Editorial. *Imperial Valley Press,* October 5.

Shupe, Steven J., Gary D. Weatherford, and Elizabeth Checchio. 1989. "Western Water Rights: The Era of Reallocation." *Natural Resources Journal* 29, No. 2 (spring): 413–434.

Smith, Rodney T., and Roger J. Vaughan. 1987. "Taking Water to Market." *Civil Engineering* (March): 70–73.

———, eds. 1993. "The Forgotten Economics of Water Trades." *Water Strategist* 7, No. 1 (April): 1–16.

Sorensen, Tom. 1989. "Water Should Be Rationed the Democratic Way—by Price." *San Jose Business Journal,* April 3.

Stanworth, Garn T. 1995. President, Stanworth Crop Consultants, Palo Verde Valley. Personal interview, March 31.

Stavins, Robert N. 1983. *Trading Investments for Water.* Berkeley, Calif.: EDF.

Summers, Joseph. 1996. President, Summers Engineering, Inc., Hanford, California. Personal interview, February 13.

U.S. Department of Commerce. 1998. "Regional Economic Information System." Economics and Statistics Administration, Bureau of Economic Analysis, Regional Economic Measurement Division. CD-ROM. Washington, D.C.

U.S. Geological Survey. 1986. National Water Summaries 1985—Hydrologic Events and Surface-Water Resources.

U.S. Geological Survey Water-Surface Paper No. 2300. Reston, Va.: U.S. Geological Survey.

Vaux, H. A., Jr., and R. Howitt. 1984. "Managing Water Scarcity: An Evaluation of Interregional Transfers." *Water Resources Research* 20, No. 7:785–792.

Villarejo, Don. 1996. "93640 at Risk: Farmers, Workers, and Townspeople in an Era of Water Uncertainty." Davis: California Institute for Rural Studies.

Walston, Roderick E. 1989. "The Public Trust Doctrine in the Water Rights Context." *Natural Resources Journal* 29, No. 2 (spring): 585–592.

Warchol, Richard. 1992. "Yolo Wants Compensation for Jobs Lost During Water Sales." *Davis Enterprise,* January 8, A1.

Water Gazette. n.d. "Federal Figures Show CV Crops Among Most Valuable." *Water Gazette* 29, No. 2.

Weatherford, Gary D. 1994. "Liquid Transfers." *The Recorder,* January 11, 7.

Weber, Gregory S. 1994. "The Role of Environmental Law in the California Water Allocation and Use System: An Overview." *Pacific Law Journal* 25:907–972.

Wheeler, Douglas P. 1992. "Planning for California's Water Future." *The Hayward Daily Review* (April 5): p. A-11.

Index

Model Water Transfer Act for
California, 29–31, 46, 159
Seven-Party Agreement of 1931, 71–73
Williamson Act, 125
Legislative efforts put into creating water
markets, 5–6, 43–45, 153–62
Lemons, 97
Lessons from California's water transfer
experiences:
cooperation in the West, water
marketing and rural–urban, 148–50
design recommendation for water
markets, 141–48
Devil's Den Water District–Castaic Lake
Water Agency combined land
purchase and water transfer, 129–32
Palo Verde Irrigation District–
Metropolitan Water District land
fallowing, 114–15
short-term markets and long-term
transfers, 49–50, 134–41
why change California's water policy
and why change to a market, 133–34
"Letting Market Forces Settle Who Gets
Water" (Nolte), 157
Lettuce, 58, 97
Liability of execution, 44
Liberty, relationship between markets and,
26
Lining canals, 85–86
Location of water use, shifts in, 5
Long-term and short-term water markets,
similarities/differences in, 60–62,
134–41, 177
see also Devil's Den Water District–
Castaic Lake Water Agency combined
land purchase and water transfer;
Imperial Irrigation District–
Metropolitan Water District water
conservation agreement; Market-
based water reallocation; Palo Verde
Irrigation District–Metropolitan
Water District land fallowing
Los Angeles, 50, 125, 148–49
Los Angeles County Waterworks District
No. 36, 122
Los Angeles Department of Water and
Power (LADWP), 69, 131
"Lost Amid the Water People" (King),
xiii

Machado, Mike, 156

Magnitude and duration of the transfer,
139
Manage, the right to, 44
"Managing Water Scarcity: An Evaluation
of Interregional Transfers" (Vaux &
Howitt), 157
Mann, John, Jr., 43
Market-based water reallocation:
advocates of, xviii, 25–27, 33–34,
46–47
analysis of water markets, institutional,
xx–xxii
Arizona, 10
change, market's ability to adapt to,
45–46
conclusions, 31–32
economic efficiency, 25–26
evaluating the feasibility and
appropriateness of, 169–73, 176–77
exploratory and experimental phase of
policy development, 15–16
freedom, enhanced individual, 26
good idea or not, 46–47
historical evolution of water rights,
38–45
introduction of, xviii
limitations of and problems with, 34–35
market institution, the, 21–25
Model Water Transfer Act for
California, 29–31
opponents of, xviii–xix
policy options, additional, 26–27
political nature of, 35–37
Texas, 12
where is the market for water rights, xx
see also Devil's Den Water District–
Castaic Lake Water Agency combined
land purchase and water transfer;
Imperial Irrigation District–
Metropolitan Water District water
conservation agreement; Palo Verde
Irrigation District–Metropolitan
Water District land fallowing
"Marketing Idea Key to Ending Water
Wars" (Elias), 159
Markets:
forms that markets take, 27–29
questioning/examining, xi
socialist–capitalist polarity, x–xi
see also Efficiency, market
Maughan, W. Donald, 83
Measurement Committee, 106

APPENDIX 3

Proposals for Market-Based Reallocation: Evaluating Their Feasibility and Appropriateness

"Human beings owe each other help to distinguish the better from the worse."

—John Stuart Mill, *On Liberty* (1859)

Most evaluations of market mechanisms center on the familiar dichotomy of efficiency versus equity. An efficient allocation results in all goods ending up in the hands of those who value them most, given everyone's initial individual wealth. If one party can make better use of a good than the party who currently owns it, the two parties will strike a deal and transfer the good. Efficient markets reduce waste because all goods either are put to their best economic use by the current owner or are sold to somebody else who will do so. An economically efficient system attaches no preference to the direction of trading (i.e., who and where are the buyers and who and where are the sellers). Traders are faceless; what matters is that if potential gains are to be had from trade, the trade is done.

Equity refers to a society's sense of fairness with respect to price-based allocation of goods and services. On the positive side, the vision of a nondiscriminatory choice of trading partners has been a leading moral argument in favor of pro-market reforms for two centuries. The market mechanism, with its promises of fairness and general prosperity for all participants, draws moral validity from its perceived neutrality with respect to who ends up with money, who ends up with goods, and in what combinations.

But from an equity perspective, markets also have downsides. When price is the arbiter of who gets what, the rich have an advantage over the poor. For some goods and services, such as provision of potable water for human use, pure market-based allocation assaults one's sense of equity. With a "pure" market, the poorest in a society would probably be priced out of potable-water use, with resulting negative effects on human health and quality of life. Even in bulk purchases of untreated water, poorer regions might suffer setbacks if their water is transferred away. Thus, the typical discussion of markets versus other forms of resource allocation pits the relative superiority of a market in terms of its efficiency against the relative superiority of the proposed alternative in terms of its equity.

Equity and efficiency are not the only standards by which market mechanisms can be judged. In fact, the two are similar in a critical sense: both are concerned with how appropriate a market mechanism is for a given task in a given context. That is, both arguments are structured in the same way: "This proposed market mechanism is efficient or equitable, and that is appropriate." The broader underlying standards for evaluating the effectiveness of a reallocation mechanism can therefore be labeled *appropriateness*. A second, similar standard can be labeled *feasibility*. Efficiency and equity, along with other considerations, would fall under the appropriateness standard; other considerations would fall under the feasibility standard.

Feasibility refers to the instrumental nature of the proposal: to what extent does it achieve the policy goal that motivates its consideration? In the case of water reallocation in California, the policy goal is to shift water use from agricultural applications to urban and environmental applications. The feasibility standard further includes the question of implementation: can the plan, once approved, actually be installed and made operational? One aspect of feasibility not covered concerns the matter of securing the political go-ahead (e.g., legislative votes or key signatures) to launch the program. This is because an analysis of a program's feasibility would generally precede and inform the political process that selects the program.

Appropriateness refers to the proposed policy from the perspective of social values. To what extent is the proposal, including its implementation mechanism, consistent with a society's values, narrowly or broadly construed? Different lists of criteria would be generated on the basis of the attributes of the society, the resource in question, and the reallocation goals. Equity, efficiency, respect for the individual, respect for communities, respect for nature, reduction of wasteful use, and protection from disasters are among the long list of sometimes contradictory values that should be considered in the context of reallocating California's water.

The appropriateness standard matters not just as a subject for academic conversation: the California state legislature itself has called for water transfers that are consistent with the well-being of both sending and receiving regions as well as with protection of nature. That is, the legislature wants transfers that are appropriate for the state given its current needs and values. (The quote is attached to note 2 on page 154.)

The categories of feasibility and appropriateness are not completely distinct. That is, it is possible that a proposal could be infeasible because it is inappropriate. Parties who perceive that they would be unfairly treated by a proposal may have the legal or administrative means to block its implementation. So although feasibility and appropriateness are separate standards in terms of their focus (instrumentality versus consistency with values), they also can be mutually reinforcing.

Figure A.3.1 is a matrix with five boxes in which separate evaluations of a market mechanism (or other policy implementation option) could be made. On the left of the matrix are two standards of comparison, feasibility and appropriateness. Along the top of the matrix are three modes of comparison:

1. How does the proposal compare with other proposals intended to accomplish the same goal?

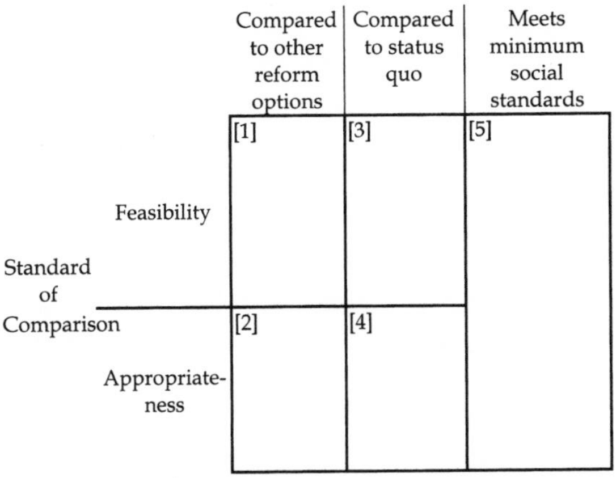

	Compared to other reform options	Compared to status quo	Meets minimum social standards
Feasibility	[1]	[3]	[5]
Appropriateness	[2]	[4]	

Standard of Comparison

FIGURE A.3.1
Matrix for Evaluating Institutional-Reform Proposals

2. How does the proposal compare with the existing set of institutions, or the status quo? Existing ways of doing things have the built-in advantage of having had practical experience and having adapted over time to specific, complex circumstances that might not be apparent to program designers.

3. Does the proposal meet minimum social standards? Even if it is shown to be superior to all other alternatives and the status quo, the proposal still may not be "good enough." If pursuing a not-good-enough option forecloses the opportunity to implement a future, superior option (perhaps because of a fixed budget or an irreversible adverse effect associated with implementation), a decision maker may be better off not adopting the otherwise superior existing option and waiting for a better one to be developed. A proposal to build a major new dam on an ecologically significant waterway might fall into this category.

Each of the boxes in figure A.3.1 asks for a separate evaluation. Boxes 1 and 3 are concerned with comparative feasibility tests. For example, one can test for comparative feasibility of a water-market proposal by examining how costly it would be to transact. How many legal steps need to be taken to complete the deal? How costly are the steps? How time-consuming are the steps? Who is responsible for enforcing the agreement? What enforcement mechanisms are available? How flexible are the transfer rules in terms of unexpected future contingencies?

An analysis of boxes 2 and 4, which are concerned with comparative appropriateness, might include an examination of the extent to which proposals could injure income equality, access to future economic opportunity, sustainable development, or biodiversity. The extent to which a proposal reduces existing wasteful uses of water, as well as other aspects of the efficiency question, could also be explored. One might further examine whether and to what extent proposals are consistent with long-term trends in legal interpretation or precedents, which could stand as a proxy for a society's evolving cultural relationship with water.

An analysis pertinent to box 5 could include an examination of comparative costs for achieving other similar policy goals, potential adverse effects on public health, or irreversible adverse environmental effects such as species extinctions or landscape reconfiguration.

The value of this matrix is that it elaborates and clarifies the scope and comparative nature of the proposal-evaluation process. It also sorts out and makes space for both normative and positive criteria, both of which are necessary for a thorough evaluation. The boundaries around each box are not impermeable. Indeed, the discussion of the evolution of property rights in chapter 3 contains perspectives that would fall into all three appropriateness boxes (2, 4, and 5).

The use of figure A.3.1 is not likely to yield an unequivocal, incontestable recommendation. Simply in terms of a binary better-or-worse ranking, there are thirty-two possible outcomes for any proposal, ranging from it being consistently superior to a single alternative proposal and the status quo to its being consistently inferior. Certainly, the closer a proposal is to arriving at a consistently superior or a consistently inferior result, the easier it is to justify a recommendation consistent with the result. Setting aside extreme results, one would also have to take into account the degree of superiority or inferiority of the various comparative categories. If a proposal is found to be substantially inferior to the status quo in terms of either appropriateness or feasibility, that could stand as grounds to cease further analysis and move on to an analysis of other existing proposals or formulation of new ones. Once formed, new options can then be subjected to the same rigorous analysis.

APPENDIX 4

Farmer Resistance to Long-Term Rural-to-Urban Water Transfers: Two Examples and an Economic Interpretation

> *Item, 1984:* Two representatives of a suburban water district located on California's rapidly expanding southern coast identify a farmer in the upper Owens Valley who wishes to sell his water rights. Conveyance of the water can be arranged through the Los Angeles Aqueduct. The representatives travel to the Owens Valley for a public hearing on the proposed transfer. On arrival at the meeting-site parking lot, the officials realize that nearly every parking space is taken by pickup trucks and all the gun racks are empty. They get back in their car and drive home.

In recent years, there has been intense social pressure in the agricultural sector against any farmer selling water to urban regions. The following are two examples. First, in 1993, the Central Valley's Areias Dairy Farm (ADF), facing bankruptcy, attempted to become the first farm to sell its water rights according to the rules of the 1992 Central Valley Project Improvement Act. The agreement reportedly called for the Metropolitan Water District of Southern California (MWD) to pay ADF $175 per acre-foot (af) for about 35,000 af of water over a period of fifteen years. MWD would be able to purchase as much as 5,000 af/year in each of any seven years it chose; ADF would receive as much as $6.1 million. Other reports put the quantity transferred at a slightly smaller amount. Two public hearings in August 1994 near the farm brought out well more than 1,000 community members, nearly all of whom opposed the trade, citing its potential adverse effects on the local community and economy. The Merced County Farm Bureau also came out against the proposed transfer, as did the area's irrigation district, the Central California Irrigation District, representatives of which said they wished they had been consulted in advance of the agreement. ADF and MWD battled for months to implement the agreement. A commissioned study pointed out that ADF occupied only 1 percent of the total land in its irrigation district, and the transferred water represented only 0.75 percent of the district's total normal-year allocation. The farm's production contributed only 0.5 percent of the county's economic output. However, opposition from the farm sector was enough not only to scuttle the trade (with the farm eventually declaring bankruptcy) but also to deal a temporary blow to the political career of Rusty Areias, a son of the farm's owner who served as state assembly representative from nearby Salinas, a city devoted to the farming industry. Areias stood by his family's right to transfer water rights. Although many other issues weighed in during the subsequent 1996 campaign, Areias eventually lost a tight race.

The second recent example of local opposition to a long-term rural-to-urban transfer involves a water-storage agreement. In 1997, a rejuvenated effort by MWD and the Arvin-Edison Water Storage District to reach a long-term agreement to move San Joaquin River water to the southern coast encountered intense opposition from Central Valley farmers. Farmers insisted that not one drop of San Joaquin River water be transferred to urban use in southern California because that could be the first step in a trend that would threaten all agricultural users of the river.

Both of these recent cases of opposition by farmers to a water transfer are consistent with a simple economic model of farmers' attitudes toward water transfers. The following paragraphs provide at best only a partial explanation for the farmers' opposition but may shed some light on their behavior with respect to water markets.

For nearly two decades, California's odd institutional setting has included both a *lack* of a long-term market in terms of actual transactions and a perception that there *is* or *could soon be* a market, in terms of institutional structures. Today's "starting place" for a long-term water market, should one arise, finds agricultural water districts in control of about 80 percent of developed water and cities with about 15 percent.[1] Cities are demanding water and can afford to pay handsomely for it. Today's lack of a water market has created a supply bottleneck, and as a result, purchase prices offered by cities are high: well over $100/af. Indeed, permanent transfers of State Water Project (SWP) entitlements were reportedly selling for $1,000/af in late 1997, with subsequent annual payments of $60 to $100/af delivered added to the agreement.

But with so much *potential supply* available to a market, should one arise, it is likely that once market transfers begin, the actual market-clearing price will be much lower than what is now being quoted. In 1995, an MWD employee in the water-marketing group privately described the agency's buying strategy as one of offering high prices at first to bring the first few farmers to the table and close some deals and then lowering the offered price significantly as trading takes off and offers to sell come pouring in.

Against this backdrop, consider one possible strategy of an individual farmer. He or she wants either to be the first to sell or to have no one else ever sell. Even if just a few other farmers sell water to cities first, the value of his or her water rights could drop precipitously—perhaps by more than 50 percent. A simple and conservative calculation suggests the economic risk to the state's farm economy if a handful of long-term rural-to-urban transfers actually take place. Assume that the value of a farmer's water rights is part of the capital value of the farm and that roughly 10 million acre-feet (maf) of water is *available* for rural-to-urban transfer, even though nowhere near that amount could ever actually be transferred. If, with the advent of a water market, the market-clearing price (excluding conveyance) for permanent water rights drops by $500/af (say, from $1,000/af to $500/af), then the perceived capitalized value of the farming sector in regions with transferable water rights could drop by $5 billion in a matter of weeks. At a minimum, this loss of value would erode the rural sector's creditworthiness, making borrowing more costly and requiring higher profit margins. This in turn would make farming in California less competitive with that in other farming regions. From this perspective, the advent of long-term water marketing presents an economic threat to large portions of California's rural economy.

If a long-term market is to arise, either it must be organized in a way that does not threaten the capital value of the state's farming sector due to the potential statewide drop in the value of agricultural water, it must be organized in a way that provides a substitute for the lost value, or it must be phased in in a way that allows farmers and rural regions to adapt effectively to the loss of capital value. The latter will most likely occur.

Note

1. The remaining 5 percent is used by federal and state governments, Indian tribes, and others.

Source Materials

Much of this book is based on my doctoral dissertation (Haddad 1996). The dissertation, as one might expect, is thick with references. To enhance the flow and readability of this book, I pared the references down considerably. If scholars or other parties are interested in pursuing further the points made here, they may peruse the dissertation, a document that is more detailed but less user-friendly. The dissertation is available at the main library of the University of California, Berkeley.

Adams, Rev. Bobby. 1988. Advertisement. *Imperial Valley Press,* December 25.

Arax, Mark. 1995. "Effort to Link Growth, Water Sparks Battle." *Los Angeles Times,* August 14.

Bakken, Gordon Morris. 1983. *The Development of Law on the Rocky Mountain Frontier.* Westport, Conn.: Greenwood Press.

Bay Area Economic Forum. 1991. *Using Water Better: A Market-Based Approach to California's Water Crisis.* October. San Francisco: Bay Area Economic Forum.

Becker, Lawrence C. 1977. *Property Rights: Philosophic Foundations.* London: Routledge & Kegan Paul.

Boronkay, Carl, and Timothy Quinn. 1993. "The Central Valley Project Improvement Act: An Urban Perspective." *San Joaquin Agricultural Law Review* 3:57–64.

Boronkay, Carl, and Charles Shreves. 1989. "Water Trades Can Help Meet Future Urban Needs." *EDF Letter* 20, No. 2 (May): 7.

Calexico Chronicle. 1989. "Secretary Hodel Applauds California Water Conservation Agreement." *Calexico Chronicle,* January 26.

Chen, Ingfei. 1991. "Water Department Gets OK to Buy on Open Market." *San Francisco Chronicle,* January 30, A-3.

Coase, Ronald. 1937. "The Nature of the Firm," *Economica N.S.* 4:386–405. In *Readings in Price Theory,* ed. G. J. Stigler and K. E. Boulding. Homewood, Ill.: Richard D. Irwin.

Conniff, Richard. 1993. "California: Desert in Disguise." *Water: The Power, Promise, and Turmoil of North America's Fresh Water,* National Geographic Magazine Special Edition, pp. 38–53.

Cooter, Robert, and Thomas Ulen. 1996. *Law and Economics.* Reading, Mass.: Addison-Wesley.

Cox, Don. 1989. "Positive Move." Editorial. *Imperial Valley Press,* January 17.

Daly, Herman E. 1996. *Beyond Growth: The Economics of Sustainable Development.* Boston: Beacon Press.

Devil's Den–Castaic Lake Authority. 1988. *Environmental Effects Outside the Castaic Lake Water Agency Service Area Resulting from the Potential Transfer of All or Part of the Contractual Water Entitlement of Devil's Den Water District to the Castaic Lake Water Agency.* Final Environmental Impact Report. October. State Clearinghouse No. 88042504. Newhall, Calif.

Dimmitt, Arnold K. 1994. "The Imperial Irrigation District/Metropolitan Water District Water Conservation Program: A Case Study." Unpublished manuscript.

Dixon, Lloyd S., Nancy Y. Moore, and Susan W. Schecter. 1993. "California's 1991 Drought Water Bank: Economic Impacts in the Selling Regions." Report prepared by the Rand Corporation for the California Department of Water Resources.

DWR (California Department of Water Resources). 1986. *A Catalog of Water Transfer Proposals.* Draft. September. Sacramento, Calif.: DWR, Water Transfers Committee.

———. 1990. *A Catalog of Water Transfer Proposals—an Update.* Draft. December. Sacramento, Calif.: DWR, Division of Local Assistance.

———. 1992. *Weekly Update on Drought Conditions.* March 13. Sacramento, Calif.: DWR.

———. 1998a. *California Water Plan Update: Bulletin 160-98.* Executive summary. Sacramento, Calif.: DWR.

———. 1998b. *California Water Plan Update: Bulletin 160-98.* Vol. 1. Sacramento, Calif.: DWR.

Elias, Tom. 1995. "Marketing Idea Key to Ending Water Wars." *Torrance Daily Breeze.* March 6, B-2.

Environmental Defense Fund. 1989. "California Agencies Sign Historic Water Swap." *EDF Letter* 20, No. 2 (May): p. 1.

Fallowing Agreement. 1992. Agreement for Land Fallowing in the Palo Verde Irrigation District ("Fallowing Agreement"). Agreement between the Metropolitan Water District of Southern California and the Palo Verde Irrigation District.

Gardner, B. Delworth. 1987. "Removing Impediments to Water Markets." *Journal of Soil and Water Conservation* (November–December): 384–388.

Garner, Eric L., Michelle Ouellette, and Richard L. Sharff Jr. 1994. "Institutional Reforms in California Groundwater Law." *Pacific Law Journal* 25:1021–1052.

Gleick, Peter H., Penn Loh, Santos V. Gomez, and Jason Morrison. 1995. *California Water 2020: A Sustainable Vision.* May. Oakland, Calif.: Pacific Institute for Studies in Development, Environment, and Security.

Great Western Research, Inc. 1995. "Palo Verde Test Land Fallowing Program, August 1, 1992–July 31, 1994, Final Report." Vol. 1. August. Report prepared for the Metropolitan Water District of Southern California.

Griffin, Ronald C., and John R. Ellis. 1997. "Water Marketing in Texas." Draft. July 15. College Station: Texas A&M University, Department of Agricultural Economics.

Grover, George S., and John F. Mann Jr. 1991. "*Acton v. Blundell* Revisited: 'Property' in California Groundwater." *Western State University Law Review* 18:589–598.

Haddad, Brent M. 1996. "Evaluating the Market Niche: Why Long-Term Rural-to-Urban Inter-regional Markets for Water Have Not Formed in California." Ph.D. diss., University of California, Berkeley.

———. 1997. *Putting Markets to Work: The Design and Use of Marketable Permits and Obligations.* Working Paper No. 19. Paris: Organization for Economic Cooperation and Development, Service de la Gestion Publique.

Hayek, F. A. 1945. "The Use of Knowledge in Society." *American Economic Review* 35, No. 4:519–530.

Hof, Robert D. 1992. "California's Next Cash Crop May Soon Be . . . Water?" *Business Week* (March 2): 76–78.

Honoré, A. M. 1961. "Ownership." Pp. 107–147 in *Oxford Essays in Jurisprudence: A Collaborative Work,* ed. A. G. Guest. London: Oxford University Press.

Howitt, Richard, Nancy Moore, and Rodney T. Smith. 1992. "A Retrospective on California's 1991 Emergency Drought Water Bank." March. Report prepared for the California Department of Water Resources.

Hull, Robert. 1996. Farm manager, Devil's Den Water District, retired. Personal interview, Corcoran, California, February 13.

Hyduke, Jeanette. 1991. "MWD Seeks Farm Land for Water." *Riverside Press-Enterprise,* December 7.

Kamand, Fadi. 1995. Engineer, Metropolitan Water District of Southern California. Personal interview, Los Angeles, March 6.

Kaplan, Karen. 1991. "State Starts Water-System Reform: Five-Year Drought Has Demonstrated Marketing Is the Way to Go." *San Diego Union,* July 21.

Kay, Jane. 1994. "Big Businesses Want Secure Water Supply." *San Francisco Examiner,* July 21.

Lapin, Lisa. 1991. "'Perestroika for Water': Drought Inspires Revolution." *San Jose Mercury News,* February 3.

MacDonald, Clyde. 1993. "Water Supply: A New Era for a Scarce Resource." Chap. 6 in *California's Threatened Environment,* ed. Tim Palmer. Washington. D.C.: Island Press.

Morris, Willy. 1989. "Water Swap Finalized After Five Years of Talks." *Imperial Valley Press,* January 1.

MWD (Metropolitan Water District of Southern California). 1986. "Metropolitan's Detailed Comments on the Draft Environmental Impact Report—Proposed Imperial Irrigation District Water Conservation Program and Initial Water

Transfer." June 13. Reprinted in *Final Environmental Impact Report,* October 1986, 4–87 ff. Los Angeles: MWD.

———. 1989. *Water Conservation Agreement between the Metropolitan Water District of Southern California and the Imperial Irrigation District.* Contains copies of the Water Conservation Agreement, Approval Agreement, and Agreement to Supplement Approval Agreement. December. Los Angeles: MWD.

———. 1990. *The Regional Urban Water Management Plan for the Metropolitan Water District of Southern California.* November. Los Angeles: MWD.

———. 1994. "Integrated Resource Plan: Phase 1 Report." Final draft. November 28. Los Angeles: MWD.

Olmstead, Janis. 1997. "Emerging Markets in Water: A Comparative Institutional Analysis of the Central Valley Project and the Colorado–Big Thompson." Working paper. September 22. Berkeley: University of California, Department of Agricultural and Resource Economics.

Porgens, Patrick, and Lloyd G. Carter. 1992. "How to Get Rich off Public Water." *Sacramento Bee,* February 16.

Producers Cotton Oil Company. n.d. Company brochure, circa 1987. Memphis, Tenn.: Producers Cotton Oil Company.

Program Agreement. 1992. "Agreement for the Implementation of a Test Land Fallowing Program and Use of Saved Water (Program Agreement)." Agreement between the Metropolitan Water District of Southern California and the Palo Verde Irrigation District. May 29.

PVID (Palo Verde Irrigation District). 1994. "Palo Verde Irrigation District, Blythe, California." Informational handout. August. Blythe, Calif.: PVID.

Reinhold, Robert. 1992. "Farmers in West May Sell Something More Valuable Than Any Crop: Water." *New York Times,* April 6.

Rice, P. A. 1990. "Letter Urged Telling Sea's Problems in Conservation." *Imperial Valley Press,* May 25, A-1.

Robie, Ronald, and Russel Kletzing. 1979. "Area of Origin Statutes—the California Experience." *Idaho Law Review* 15:419–441.

Rosegrant, Mark W. 1995. "Water Transfers in California: Potentials and Constraints." *Water International* 20:72–87.

Rosenberg, Richard M. 1995. "A New Era in California Water: The Business Perspective." Speech delivered to the Water Education Foundation, March 30. Sacramento, Calif.

Sagehorn, Robert. 1995. General manager, Castaic Lake Water Agency. Personal interview, Castaic Lake, California, June 12.

San Francisco Chronicle. 1991. "Emergency Water Bank Has a Surplus." *San Francisco Chronicle,* November 19.

Sax, Joseph L. 1990. "The Constitution, Property Rights, and the Future of Water Law." *University of Colorado Law Review* 61:257–282.

Sax, Joseph L., Robert H. Abrams, and Barton H. Thompson Jr. 1991. *Legal Control of Water Resources.* 2nd ed. St. Paul, Minn.: West.

Schulz, Clifford W., and Gregory S. Weber. 1988. "Changing Judicial Attitudes towards Property Rights in California Water Resources: From Vested Rights to Utilitarian Reallocations." *Pacific Law Journal* 19:1031–1110.

Severns, Marjorie. 1989. "Voice of the People." Editorial. *Imperial Valley Press,* October 5.

Shupe, Steven J., Gary D. Weatherford, and Elizabeth Checchio. 1989. "Western Water Rights: The Era of Reallocation." *Natural Resources Journal* 29, No. 2 (spring): 413–434.

Smith, Rodney T., and Roger J. Vaughan. 1987. "Taking Water to Market." *Civil Engineering* (March): 70–73.

————, eds. 1993. "The Forgotten Economics of Water Trades." *Water Strategist* 7, No. 1 (April): 1–16.

Sorensen, Tom. 1989. "Water Should Be Rationed the Democratic Way—by Price." *San Jose Business Journal,* April 3.

Stanworth, Garn T. 1995. President, Stanworth Crop Consultants, Palo Verde Valley. Personal interview, March 31.

Stavins, Robert N. 1983. *Trading Investments for Water.* Berkeley, Calif.: EDF.

Summers, Joseph. 1996. President, Summers Engineering, Inc., Hanford, California. Personal interview, February 13.

U.S. Department of Commerce. 1998. "Regional Economic Information System." Economics and Statistics Administration, Bureau of Economic Analysis, Regional Economic Measurement Division. CD-ROM. Washington, D.C.

U.S. Geological Survey. 1986. National Water Summaries 1985—Hydrologic Events and Surface-Water Resources.

U.S. Geological Survey Water-Surface Paper No. 2300. Reston, Va.: U.S. Geological Survey.

Vaux, H. A., Jr., and R. Howitt. 1984. "Managing Water Scarcity: An Evaluation of Interregional Transfers." *Water Resources Research* 20, No. 7:785–792.

Villarejo, Don. 1996. "93640 at Risk: Farmers, Workers, and Townspeople in an Era of Water Uncertainty." Davis: California Institute for Rural Studies.

Walston, Roderick E. 1989. "The Public Trust Doctrine in the Water Rights Context." *Natural Resources Journal* 29, No. 2 (spring): 585–592.

Warchol, Richard. 1992. "Yolo Wants Compensation for Jobs Lost During Water Sales." *Davis Enterprise,* January 8, A1.

Water Gazette. n.d. "Federal Figures Show CV Crops Among Most Valuable." *Water Gazette* 29, No. 2.

Weatherford, Gary D. 1994. "Liquid Transfers." *The Recorder,* January 11, 7.

Weber, Gregory S. 1994. "The Role of Environmental Law in the California Water Allocation and Use System: An Overview." *Pacific Law Journal* 25:907–972.

Wheeler, Douglas P. 1992. "Planning for California's Water Future." *The Hayward Daily Review* (April 5): p. A-11.

Index